FLORA & FAUNA HANDBOOK NO. 8

# MAYFLIES OF THE WORLD

# A Catalog of the Family and Genus Group Taxa

## (INSECTA: EPHEMEROPTERA)

UNIVERSAL BOOK
SERVICES

DR. W. BACKHUYS
OEGSTGEEST

# Flora & Fauna Handbooks

This series of handbooks provides for the publication of book length working tools useful to systematics for the identification of specimens, as a source of ecological and life history information, and for information about the classification of plant and animal taxa. Each book is sequentially numbered, starting with Handbook No. 1, as a continuing series. The books are available on a standing order basis, or singly.

Each book treats a single biological group of organisms (e.g., family, subfamily, single genus, etc.) or the ecology of certain organisms or certain regions. Catalogs and checklists of groups not covered in other series are included in this series.

The books are complete by themselves, not a continuation or supplement to an existing work, or requiring another work in order to use this one. The books are comprehensive, and therefore, of general interest.

Books in this series to date are:

Handbook No. 1 - THE SEDGE MOTHS, by John B. Heppner

Handbook No. 2 - INSECTS AND PLANTS: Parallel Evolution and Adaptations, by Pierre Jolivet

Handbook No. 3 - THE POTATO BEETLES, by Richard L. Jacques, Jr.

Handbook No. 4 - THE PREDACEOUS MIDGES OF THE WORLD, by Willis W. Wirth and William L. Grogan, Jr.

Handbook No. 5 - A CATALOG OF THE NEOTROPICAL COLLEMBOLA, by José A. Mari Mutt and Peter F. Bellinger

Handbook No. 6 - THE ENDANGERED ANIMALS OF THAILAND, by Stephen R. Humphries and James R. Bain

Handbook No. 7 - A REVIEW OF THE GENERA OF NEW WORLD MYMARIDAE, by Carl M. Yoshimoto

Handbook No. 8 - MAYFLIES OF THE WORLD, A Catalog of the Family and Genus Group Taxa, by Michael D. Hubbard

FLORA & FAUNA HANDBOOK NO. 8

# MAYFLIES OF THE WORLD

# A Catalog of the Family and Genus Group Taxa

## (INSECTA: EPHEMEROPTERA)

by

Michael D. Hubbard
Florida A & M University
Tallahassee, Florida

SANDHILL CRANE PRESS, INC.
Gainesville, Florida
1990

Editor: Ross H. Arnett, Jr.

ISBN 1-877743-06-2

A FLORA & FAUNA HANDBOOK

**Library of Congress Cataloging in Publication Data:**

Hubbard, Michael D.
Mayflies of the world : a catalog of the family and genus group taxa :
(Insecta : Ephemeroptera) / by Michael D. Hubbard.
      p.   cm. -- (Flora & Fauna handbook ; no. 8)
    Includes bibliogrpahical references.
    ISBN 1-877743-06-2 (paper) : $24.95
    1. Mayflies--classification.  I. Title.  II. Series.
QL 505.H83   1990
595.7'34'012--dc20          90-48456
                CIP

Manufactured in the United States of America

*Published jointly by*:
The Sandhill Crane Press, Inc.
Mail: P. O. Box 1702-100B
Office: 2406 N.W. 47th Terrace
Gainesville, FL 32606, USA

    *And*:
Universal Book Services
Dr. W. Backhuys
Warmonderweg 80
2341 KZ Oegstgeest, The Netherlands

# TABLE OF CONTENTS

# INTRODUCTION

This catalog lists all family-group (family, subfamily, tribe, subtribe) and genus-group (generic or subgeneric) names which have been proposed for the Ephemeroptera, both Recent and fossil. It is divided into four parts. The first part illustrates the hierarchical classification of the Ephemeroptera. The second part is an alphabetical list of all family-group names which have been proposed for Ephemeroptera. The third part consists of an alphabetical catalog of all genus-group names which have been proposed for Ephemeroptera. The fourth section consists of full references to the literature cited in the catalog.

There are presently known 371 valid genera of Ephemeroptera, of which 61 are known only from fossils. This contrasts with approximately 231 valid genera known in 1976 (Hubbard & Peters 1976).

The classification scheme used in this publication is based largely on that proposed by Landa & Soldán (1985) modified slightly by my own somewhat idiosyncratic views of classification. In the hierarchical classification scheme each valid genus-group taxon is listed in the appropriate systematic position in the hierarchy, indicating the family-group taxa to which it belongs. If there are synonyms for any of the genus-group taxa, they are listed following the appropriate name. The geographic distribution (or in the case of fossil taxa, indicated by a preceeding "†", the geologic age) of each valid genus-group taxon is given in parentheses following each name.

I have tried to follow closely the International Rules of Zoological Nomenclature in the listing and discussion of the taxa in this publication.

In the third part of this catalog, the alphabetical catalog of genus-group names, the main entry for each valid name is **bold-faced**. If the genus-group name is invalid (i.e., not available or a junior synonym) the entry is *italicized*. The main entry for each genus-group name also lists the author(s) of the name, with the date of publication of the name, and the starting page number of the original description or proposal of the name which made the name taxonomically available. The type-species of each genus-group taxon is listed and the method of designation of the type-species is given (whether original designation, monotypy, subsequent designation, or objective synonymy)

For valid names, the family-group categories to which each taxon belongs is given, along with any synonyms of the genus-group taxon. For invalid names, the reason for the invalidity or the senior synonym is given. Where a senior synonym exists, a reference to a declaration or discussion of the synonymy is included. Where appropriate, further notes concerning the nomenclature and systematics of the taxon are listed in a following paragraph.

The fourth section, Literature Cited, lists, alphabetically by author, the full bibliographic reference to all papers cited in the text.

This paper is one in a continuing series of catalogs of the Ephemeroptera (cf. Hubbard & Peters 1978, Hubbard & Pescador 1978, Hubbard 1979b, 1982a, 1982b, 1986, 1987, Hubbard & Savage 1981).

## ACKNOWLEDGEMENTS

In preparing this paper I drew heavily on the literature collection and Ephemeroptera index files of Dr William L. Peters and Mrs Janice G. Peters whose help I gratefully acknowledge. Dr Eduardo Dominguez, Dr R. Wills Flowers, and Dr Manuel L. Pescador gave me encouragement, suggestions, and the benefit of their knowledge. Dr Jean-Marc Elouard and Dr Michel Sartori were tireless in their encouragement of the publication of this project. I must also thank all the mayfly workers who sent me copies of their papers over the years.

# HIERARCHICAL CLASSIFICATION SCHEME OF THE ORDER EPHEMEROPTERA

In this section each valid genus and subgenus is listed in the hierarchical classification scheme along with any synonyms. For Recent genera, the known geographical distribution of the genus is given (Af = Africa; As = Asia; Aus = Austalia; E = Europe; Mad = Madagascar; NA = North America; NZ = New Zealand; SA = South America [CA = Central America only]). For genera known only from fossils, marked with a "†", the geological age is given.

†Protoephemeroptera
  †Triplosoboidea
    †Triplosobidae
          †**Triplosoba** Handlirsch [= **Blanchardia** Brongniart] (Carboniferous)

†Permoplectoptera
  †Syntonopteroidea
    †Bojophlebiidae
          †**Bojophlebia** Kukalová-Peck (Carboniferous)

    †Syntonopteridae
          †**Lithoneura** Carpenter (Carboniferous)
          †**Syntonoptera** Handlirsch (Carboniferous)

  †Protereismatoidea
    †Jarmilidae
          †**Jarmila** Demoulin (Permian)

    †Mesoplectopteridae
          †**Mesoplectopteron** Handlirsch (Triassic, Permian)

    †Misthodotidae
          †**Misthodotes** Sellards [= **Dromeus** Sellards; = **Eudoter** Tillyard] (Permian)

    †Oboriphlebiidae
          †**Oboriphlebia** Hubbard & Kukalová- Peck (Permian)

    †Protereismatidae
          †**Kukalova** Demoulin (Permian)
          †**Phthartus** Handlirsch (Permian)
          †**Protereisma** Sellards [= **Protechma** Sellards; = **Prodromus** Sellards; = **Bantiska** Sellards; = **Rekter** Sellards; =

†Mesephemeroidea
†Mesephemeridae

**Pinctodia** Sellards; = **Scopus** Sellards; = **Mecus** Sellards; = **Recter** Sellards; = **Prodromites** Cockerell; = **Loxophlebia** Martynov] (Permian)

†**Mesephemera** Handlirsch (Jurassic)
†**Palinephemera** Lin (Jurassic)
†**Palingeniopsis** Martynov (Permian)

Baetoidea
Ameletopsidae
Ameletopsinae

**Ameletopsis** Phillips (NZ)
†**Balticophlebia** Demoulin (Eocene)
**Chaquihua** Demoulin (SA)
**Mirawara** Harker (Aus)

Chiloporterinae

**Chiloporter** Lestage (SA)

Ametropodidae
Ametropodinae

**Ametropus** Albarda (As, E, NA)
†**Brevitibia** Demoulin (Eocene)

Baetidae

**Acanthiops** Waltz & McCafferty (Af)
**Acentrella** Bengtsson (Af, As, E)
**Acerpenna** Waltz & McCafferty (NA)
**Afrobaetodes** Demoulin (Af)
**Apobaetis** Day (NA)
**Baetis** Leach [= **Brachyphlebia** Westwood]
    s.g. **Baetiella** Uéno [= **Neobaetiella** Müller-Liebenau] (As)
    s.g. **Baetis** Leach (Af, As, Aus, E, Mad, NA, SA)
    s.g. **Holobaetis** Sukatskene [not available]
    s.g. **Labiobaetis** Novikova & Kluge (As)
    s.g. **Montobaetis** Kazlauskas [not available]
    s.g. **Nigrobaetis** Kazlauskas (As)
    s.g. **Rhodobaetis** Kazlauskas [not available]
    s.g. **Takobia** Novikova & Kluge (As)

s.g. **Vernobaetis** Kazlauskas [not available]
**Baetodes** Needham & Murphy (NA, SA)
**Baetopus** Keffermüller
  s.g. **Baetopus** Keffermüller (As, E)
  s.g. **Raptobaetopus** Müller-Liebenau (As, E)
**Barbaetis** Waltz & McCafferty (NA)
**Bernerius** Waltz & McCafferty (SA)
**Bungona** Harker (Aus)
**Callibaetis** Eaton [= **Neobaetis** Navás] (NA, SA)
**Camelobaetidius** Demoulin (SA)
**Centroptiloides** Lestage [= **Haplobaetis** Navás] (Af, Mad)
**Centroptilum** Eaton (Af, As, Aus, E, Mad, NA)
**Chopralla** Waltz & McCafferty (As)
**Cloeodes** Traver [= **Centroptella** Braasch & Soldán]
  s.g. **Cloeodes** Traver (As, SA, NA)
  s.g. **Notobaetis** Morihara & Edmunds (SA)
**Cloeon** Leach [= **Neocloeon** Traver; = **Austrocloeon** Barnard; = **Cloe** Burmeister; = **Cloeopsis** Eaton] (Af, As, Aus, E, Mad, NA, SA)
**Dactylobaetis** Traver & Edmunds (Na, SA)
**Diphetor** Waltz & McCafferty (NA)
**Echinobaetis** Mol (As)
**Fallceon** Waltz & McCafferty (NA)
**Guajirolus** Flowers (SA)
**Harpagobaetis** Mol (SA)
**Heterocloeon** McDunnough [= **Rheobaetis** Müller-Liebenau] (NA)
**Indobaetis** Müller-Liebenau & Morihara (As)
**Indocloeon** Müller-Liebenau (As)
**Jubabaetis** Müller-Liebenau (As)
**Libebiella** Waltz & McCafferty
  **Matsumuracloeon** Hubbard [= **Pseudocloeon** Matsumura, preoccupied] (AS)

**†Mesobaetis** Brauer, Redtenbacher, & Ganglbauer (Jurassic)
**Moribaetis** Waltz & McCafferty
    **s.g. Mayobaetis** Waltz & McCafferty (SA)
    **s.g. Moribaetis** Waltz & McCafferty (SA)
**Mutelocloeon** Gillies & Elouard (Af)
**Nesoptiloides** Demoulin (Mad)
**Ophelmatostoma** Waltz & McCafferty (Af)
**†Palaeobaetodes** Brito (Cretaceous)
**Paracloeodes** Day (NA, SA)
**Platybaetis** Müller-Liebenau (As)
**Procloeon** Bengtsson [= **Pseudocloeon** Bengtsson] (Af, As, E)
**Promatsumura** Hubbard [= **Procloeon** Matsumura, preoccupied] (As)
**Pseudocentroptiloides** Jacob
    **s.g. Psammonella** Glazaczow (As)
    **s.g. Pseudocentroptiloides** Jacob (E)
**Pseudocentroptilum** Bogoesco (E)
**Pseudocloeon** Klapálek [= **Baetiella** Uéno] (Af, As, E, Mad, NA, SA)
**Pseudopannota** Waltz & McCafferty (Af)
**Rhithrocloeon** Gillies (Af)
**Symbiocloeon** Müller-Liebenau (As)

Siphlaenigmatidae

**Siphlaenigma** Penniket (NZ)

Oniscigastridae

**Oniscigaster** McLachlan (NZ)
**Siphlonella** Needham & Murphy (SA)
**Tasmanophlebia** Tillyard [= **Tasmanophlebioides** Lestage] (Aus)

Siphlonuridae
  Acanthametropodinae

**Acanthametropus** Tshernova [= **Metreturus** Burks] (As, NA)
**†Albisca** Sinitshenkova (Mesozoic)
**Analetris** Edmunds (NA)
**Siphluriscus** Ulmer (As)
**†Stackelbergisca** Tshernova (Jurassic)

  Metretopodinae

**Metretopus** Eaton (As, E, NA)

**Siphloplecton** Clemens (NA)

Rallidentinae

**Rallidens** Penniket (NZ)

Siphlonurinae

**Ameletoides** Tillyard (Aus)
**Ameletus** Eaton [= **Chimura** Navás; = **Metreletus** Demoulin; = **Paleoameletus** Lestage] (As, E, NA)
†**Aphelophlebodes** Pierce (Miocene)
†**Australurus** Jell & Duncan (Cretaceous)
†**Baltameletus** Demoulin (Eocene)
**Dipteromimus** McLachlan [= **Dipteromimodes** Matsumura] (As)
†**Dulcimana** Jell & Duncan (Cretaceous)
**Edmundsius** Day (NA)
**Metamonius** Eaton (SA)
**Nesameletus** Tillyard (NZ)
†**Olgisca** Demoulin (Jurassic)
**Parameletus** Bengtsson [= **Eatonia** Bengtsson; = **Palmenia** Lestage; = **Potameis** Bengtsson; = **Siphlonuroides** McDunnough; = **Sparrea** Esben-Petersen] (As, E, NA)
†**Proameletus** Sinitshenkova (Cretaceous)
†**Promirara** Jell & Duncan (Cretaceous)
**Siphlonisca** Needham (As, NA)
**Siphlonurus** Eaton [= **Siphlurella** Bengtsson; = **Andromina** Navás; = **Siphlurus** Eaton] (As, E, NA)

Hexagenitoidea
†Aenigmephemeridae

†**Aenigmephemera** Tshernova (Jurassic)

†Aphelophlebodidae

†**Aphelophlebodes** Pierce (Miocene)

†Hexagenitidae

†**Ephemeropsis** Eichwald [= **Phacelobranchus** Handlirsch] (Jurassic)
†**Hexagenites** Scudder [= **Paedephemera** Handlirsch; = **Stenodicranum** Demoulin] (Jurassic)
†**Hexameropsis** Tshernova & Sinitshenkova (Lower Cretaceous)

†**Mongologenites** Sinitshenkova (Cretaceous)
†**Sibirogenites** Sinitshenkova (Jurassic)

Heptagenioidea
†Epeoromimidae

†**Epeoromimus** Tshernova (Jurassic, Cretaceous)
†**Foliomimus** Sinitshenkova (Jurassic)

Oligoneuriidae
Chromarcyinae

**Chromarcys** Navás [= **Pseudoligoneuria** Ulmer] (As)

Coloburiscinae

**Coloburiscoides** Lestage (Aus)
**Coloburiscus** Eaton [= **Coloburus** Eaton] (NZ)
†**Cronicus** Eaton (Eocene)
†**Mogzonurella** Sinitshenkova (Jurassic)
†**Mogzonurus** Sinitshenkova (Jurassic)
**Murphyella** Lestage [= **Dictyosiphlon** Lestage] (SA)
†**Siphlurites** Cockerell (Miocene)

Isonychiinae

**Isonychia** Eaton [= **Jolia** Eaton; = **Chirotonetes** Needham; = **Eatonia** Ali]
s.g. **Borisonychia** McCafferty (NA)
s.g. **Isonychia** Eaton (As, E, NA, SA)
s.g. **Prionoides** Kondratieff & Voshell (NA)

Oligoneuriinae

**Elassoneuria** Eaton
s.g. **Elassoneuria** Eaton (Af, Mad)
s.g. **Madeconeuria** Demoulin (Mad)
**Homeoneuria** Eaton
s.g. **Homeoneuria** Eaton (NA, SA)
s.g. **Notachora** Pescador & Peters (SA)
**Lachlania** Hagen [= **Alloydia** Needham; = **Neophlebia** Navás; = **Noya** Navás; = **Noyopsis** Navás] (NA, SA)
**Oligoneuria** Pictet (SA)
**Oligoneuriella** Ulmer (Af, As, E)
**Oligoneurioides** Demoulin (SA)

**Oligoneuriopsis** Crass (Af)
**Oligoneurisca** Lestage (E)
†**Protoligoneuria** Demoulin (Calcareous)
**Spaniophlebia** Eaton (SA)

Arthropleidae

**Arthroplea** Bengtsson [= **Haplogenia** Blair; = **Remipalpus** Bengtsson] (E, NA)
†**Electrogenia** Demoulin (Eocene)

Heptageniidae
Anepeorinae

**Anepeorus** McDunnough [= **Spinadis** Edmunds & Jensen] (As, NA)
**Pseudiron** McDunnough (NA)

Heptageniinae

**Acanthomola** Whiting & Lehmkuhl (NA)
**Afronurus** Lestage (Af, As, E)
**Asionurus** Braasch & Soldán (As)
**Atopopus** Eaton (As)
**Belovius** Tshernova (As)
**Bleptus** Eaton (As)
**Cinygma** Eaton (As, E, NA)
**Cinygmoides** Matsumura (As)
**Cinygmula** McDunnough (As, NA)
**Compsoneuria** Eaton [= **Compsoneuriella** Ulmer; = **Notonurus** Crass] (Af, As, Mad)
**Ecdyonurus** Eaton [= **Afghanurus** Demoulin; = **Akkarion** Flowers; = **Cinygmina** Kimmins; = **Ecdyurus** Eaton; = **Leucrocuta** Flowers; = **Nixe** Flowers; = **Notacanthurus** Tshernova; = **Paracinygmula** Bajkova] (Af, As, E, NA)
**Electrogena** Zurwerra & Tomka (E)
**Epeiron** Demoulin (As, E)
**Epeorella** Ulmer (As)
**Epeorus** Eaton
    s.g. **Epeorus** Eaton (As, E)
    s.g. **Ironopsis** Eaton (As, NA)
**Heptagenia** Walsh [= **Sigmoneuria** Demoulin]
    s.g. **Dacnogenia** Kluge (As)

s.g. **Heptagenia** Walsh (As, E, NA, SA)

s.g. **Kageronia** Matsumura (As)

**Iron** Eaton (As, E, NA, SA)

**Ironodes** Traver (NA)

**Macdunnoa** Lehmkuhl (NA)

†**Miocoenogenia** Tshernova (Oligocene)

**Ororotsia** Traver (As)

**Paegniodes** Eaton (As)

**Raptoheptagenia** Whiting & Lehmkuhl (NA)

**Rhithrogena** Eaton (Af, As, E, NA, SA [CA only])

**Rhithrogeniella** Ulmer (As)

**Stenacron** Jensen (NA)

**Stenonema** Traver

   s.g. **Maccaffertium** Bednarik (NA, SA)

   s.g. **Stenonema** Traver (NA)

†**Succinogenia** Demoulin (Eocene)

**Thalerosphyrus** Eaton [= **Ecdyonuroides** Dạng] (As)

**Trichogenia** Braasch & Soldán (As)

Leptophlebioidea
  Leptophlebiidae
    Atalophlebiinae

**Acanthophlebia** Towns (NZ)

**Adenophlebia** Eaton [= **Esbenophlebia** Lestage] (Af)

**Adenophlebiodes** Ulmer [= **Euphlebia** Crass]

   s.g. **Adenophlebiodes** Ulmer (Af)

   s.g. **Hyalophlebia** Demoulin (Af)

**Aprionyx** Barnard (Af)

**Archethraulodes** Pescador & Peters (SA)

**Arachnocolus** Towns & Peters (NZ)

**Askola** Peters (SA)

**Atalomicria** Harker (Aus)

**Atalophlebia** Eaton (As?, Aus, SA?)

**Atalophlebiodes** Phillips (Aus, NZ, SA?)

**Atopophlebia** Flowers (SA)

**Austroclima** Towns & Peters (NZ)

**Austrophlebiodes** Campbell & Suter (Aus)

**Bibulmena** Dean (Aus)
†**Blasturophlebia** Demoulin (Eocene)
**Borinquena** Traver
    s.g. **Australphlebia** Peters (SA)
    s.g. **Borinquena** Traver (SA)
**Careospina** Peters (SA)
**Castanophlebia** Barnard (Af)
**Chiusanophlebia** Uéno (As)
**Choroterpes** Eaton
    s.g. **Choroterpes** Eaton (Af, E, NA, SA)
    s.g. **Euthraulus** Barnard [= **Thraululus** Demoulin] (Af, As, E)
    s.g. **Neochoroterpes** Allen (NA, SA)
**Choroterpides** Ulmer (As)
**Celiphlebia** Peters & Peters (NC)
**Coula** Peters & Peters (NC)
**Cryophlebia** Towns & Peters (NZ)
**Cryptopenella** Gillies (As)
**Dactylophlebia** Pescador & Peters (NA, SA)
**Deleatidium** Eaton (Aus, SA?)
**Demoulinellus** Pescador & Peters (SA)
**Ecuaphlebia** Dominguez (SA)
**Edmundsula** Sivaramakrishnan (As)
**Farrodes** Peters (SA)
**Fittkaulus** Savage & Peters (SA)
**Fulleta** Navás (Af)
**Fulletomimus** Demoulin (Af)
**Garinjuga** Campbell & Suter (Aus)
**Hagenulodes** Ulmer (Mad)
**Hagenulopsis** Ulmer (SA)
**Hagenulus** Eaton (As, SA)
**Hapsiphlebia** Peters & Edmunds (SA)
**Hermanella** Needham & Murphy
    s.g. **Guayakia** Dominguez & Flowers (SA)
    s.g. **Hermanella** Needham & Murphy (SA)
**Hermanellopsis** Demoulin (SA)
**Homothraulus** Demoulin (SA)
**Hylister** Dominguez & Flowers (SA)
**Indialis** Peters & Edmunds (As)

**Isca** Gillies
   s.g. **Isca** Gillies (As)
   s.g. **Minyphlebia** Peters & Edmunds (As)
   s.g. **Tanycola** Peters & Edmunds (As)
**Isothraulus** Towns & Peters (NZ)
**Jappa** Harker (Aus)
**Kariona** Peters & Peters (NC)
**Kimminsula** Peters & Edmunds (As)
**Kirrara** Harker (Aus)
**Koorrnonga** Campbell & Suter (Aus)
**Leentvaaria** Demoulin (SA)
**Lepegenia** Peters, Peters, & Edmunds (NC)
**Lepeorus** Peters, Peters, & Edmunds (NC)
**Magallanella** Pescador & Peters (SA)
**Maheathraulus** Peters, Gillies, & Edmunds (Mad)
**Massartella** Lestage (SA)
**Massartellopsis** Demoulin (SA)
**Mauiulus** Towns & Peters (NZ)
**Megaglena** Peters & Edmunds (As)
**Meridialaris** Peters & Edmunds (SA)
**Microphlebia** Savage & Peters (SA)
**Miroculis** Edmunds
   s.g. **Atroari** Savage & Peters (SA)
   s.g. **Miroculis** Edmunds (SA)
   s.g. **Ommaethus** Savage & Peters (SA)
   s.g. **Yaruma** Savage & Peters (SA)
**Miroculitus** Savage & Peters (SA)
**Nathanella** Demoulin (As)
**Neboissophlebia** Dean (Aus)
**Needhamella** Dominguez & Flowers (SA)
**Neohagenulus** Traver (SA)
**Neozephlebia** Penniket (NZ)
**Nesophlebia** Peters & Edmunds (Mad)
**Notachalcus** Peters & Peters (NC)
**Notophlebia** Peters & Edmunds (As)
**Nousia** Navás [= **Atalonella** Needham & Murphy]
   s.g. **Australonousia** Campbell & Suter (Aus)
   s.g. **Nousia** Navás (SA)
**Nyungara** Dean (Aus)

**Ounia** Peters & Peters (NC)
**Papposa** Peters & Peters (NC)
**Peloracantha** Peters & Peters (NC)
**Penaphlebia** Peters & Edmunds (SA)
**Penniketellus** Towns & Peters (NZ)
**Perissophlebiodes** Savage [= **Perisso-phlebia** Savage] (SA)
**Petersophlebia** Demoulin (Mad)
**Petersula** Sivaramakrishnan (As)
**Polythelais** Demoulin (Mad)
**Poya** Peters & Peters (NC)
**Rhigotopus** Pescador & Peters (SA)
**Secochela** Pescador & Peters (SA)
**Simothraulopsis** Demoulin (SA)
**Simothraulus** Ulmer (As)
**Sulawesia** Peters & Edmunds (As)
**Terpides** Demoulin (SA)
**Thraulodes** Ulmer (As?, NA, SA)
**Thraulophlebia** Demoulin (Aus)
**Thraulus** Eaton [= **Masharikella** Peters & Edmunds] (Af, As, Aus, E)
**Tindea** Peters & Peters (NC)
**Traverella** Edmunds (NA, SA)
**Traverina** Peters (SA)
**Ulmeritus** Traver
    s.g. **Pseudulmeritus** Traver (SA)
    s.g. **Ulmeritoides** Traver (SA)
    s.g. **Ulmeritus** Traver (SA)
**Ulmerophlebia** Demoulin (Aus, Mad)
†**Xenophlebia** Demoulin (Eocene)
**Zephlebia** Penniket
    s.g. **Terama** Towns (NZ)
    s.g. **Zephlebia** Penniket (NZ)
Leptophlebiinae
**Calliarcys** Eaton (E)
**Dipterophlebiodes** Demoulin (As)
**Gilliesia** Peters & Edmunds (As)
**Habroleptoides** Schoenemund (Af?, E)
**Habrophlebia** Eaton
    s.g. **Habrophlebia** Eaton (E)
    s.g. **Hesperaphlebia** Peters (NA)
**Habrophlebiodes** Ulmer (As, NA, SA?)

Leptophlebia Westwood [= **Blasturus**
Eaton; = s.g. **Euphyurus** Bengtsson]
s.g. **Leptophlebia** Westwood (As, E,
NA)
**Paraleptophlebia** Lestage [= **Oligophle-
bia** Demoulin] (As, E, NA)
†Mesonetinae
†**Cretoneta** Tshernova (Cretaceous)
†**Leptoneta** Sinitshenkova (Mesozoic)
†**Mesoneta** Brauer, Redtenbacher, & Gangl-
bauer (Jurassic, Cretaceous)
Subfamily incertus
†**Lepismophlebia** Demoulin (Florissant)
Ephemeroidea
Behningiidae
†**Archeobehningia** Tshernova (Jurassic)
**Behningia** Lestage (As, E)
**Dolania** Edmunds & Traver (NA)
**Protobehningia** Tshernova (As)
Euthyplociidae
Euthyplociinae
**Campylocia** Needham & Murphy [= **Longi-
nella** Gros & Lestage] (SA)
**Euthyplocia** Eaton (As, SA)
**Mesoplocia** Demoulin (SA)
**Polyplocia** Lestage (As)
**Proboscidoplocia** Demoulin (Mad)
Exeuthyplociinae
**Afroplocia** Lestage (Af)
**Exeuthyplocia** Lestage (Af)
Ephemeridae
Ephemerinae
**Afromera** Demoulin [= **Dicrephemera** Mc-
Cafferty & Edmunds] (Af)
†**Denia** McCafferty
**Eatonica** Navás (Af, Mad)
**Eatonigenia** Ulmer (As)
**Ephemera** Linnaeus
s.g. **Aethephemera** McCafferty & Ed-
munds (As)
s.g. **Ephemera** Linnaeus [= **Nirvius**
Navás] (Af, As, E, NA)

**Hexagenia** Walsh
    s.g. **Hexagenia** Walsh (As?, NA)
    s.g. **Pseudeatonica** Spieth (SA)
**Litobrancha** McCafferty (NA)
†**Parabaetis** Haupt (Oligocene)
Ichthybotinae
**Ichthybotus** Eaton (NZ)
Palingeniidae
Palingeniinae

**Anagenesia** Eaton (As)
**Chankagenesia** Buldovskii (As)
**Cheirogenesia** Demoulin (Mad)
**Heterogenesia** Dạng (As)
†**Mesogenesia** Tshernova (Jurassic)
†**Mesopalingea** Whalley & Jarzembowski
    (Jurassic-Cretaceous)
**Mortogenesia** Lestage (As)
**Palingenia** Burmeister (Af?, As, E, SA?)
**Plethogenesia** Ulmer [= **Tritogenesia**
    Lestage] (As)
Pentageniinae
**Pentagenia** Walsh (NA)
Polymitarcyidae
Polymitarcyinae
**Ephoron** Williamson [= **Polymitarcys**
    Eaton; = **Eopolymitarcys** Tshernova]
    (Af, As, E, NA)
Campsurinae
**Campsurus** Eaton (NA, SA)
**Tortopus** Needham & Murphy (NA, SA)
Asthenopodinae
†**Asthenopodichnium** Thenius (Miocene)
**Asthenopus** Eaton [= **Asthenopodes** Ul-
    mer] (SA)
**Povilla** Navás
    s.g. **Languidipes** Hubbard (As)
    s.g. **Povilla** Navás (Af, As)
Potamanthidae

**Neopotamanthodes** Hsu (Sa)
**Neopotamanthus** Wu & You (As)
**Potamanthindus** Lestage (As)
**Potamanthodes** Ulmer (As)

**Potamanthus** Pictet [= **Eucharidis** Joly &
Joly] (Af, As, E, NA, SA?)
**Rhoenanthopsis** Ulmer (As)
**Rhoenanthus** Eaton (As)
†Torephemeridae
†**Torephemera** Sinitshenkova (Mesozoic)
Ephemerelloidea
Ephemerellidae
Ephemerellinae
Ephemerellini
Ephemerellae
**Acerella** Allen (As)
**Caudatella** Edmunds (NA)
**Caurinella** Allen (NA)
**Cincticostella** Allen
s.g. **Cincticostella** Allen [= **Asiatella**
Tshernova] (As)
s.g. **Rhionella** Allen (As)
**Crinitella** Allen & Edmunds (As)
**Drunella** Needham
s.g. **Drunella** Needham (As)
s.g. **Eatonella** Needham (NA)
s.g. **Myllonella** Allen (NA)
s.g. **Tribrochella** Allen (As)
s.g. **Unirachella** Allen (NA)
**Ephemerella** Walsh
s.g. **Ephemerella** Walsh [= **Chitono-
phora** Bengtsson] (Af, As, E, NA)
s.g. **Uracanthella** Belov (As)
**Serratella** Edmunds (As, NA)
**Teloganopsis** Ulmer (As)
**Torleya** Lestage (As, E)
†**Turfanerella** Demoulin (Jurassic)
Timpanogae
**Attenella** Edmunds [= **Attenuatella** Edm-
unds] (NA)
**Dannella** Edmunds
s.g. **Dannella** Edmunds (NA)
s.g. **Dentatella** Allen (NA)
**Eurylophella** Tiensuu [= **Melanameletus**
Tiensuu] (As, E, NA)
**Timpanoga** Needham (NA)

Vietnamellae
**Vietnamella** Tshernova (As)
Hyrtanellini
**Hyrtanella** Allen & Edmunds (As)
Melanemerellinae
**Melanemerella** Ulmer (SA)
Teloganodinae
**Ephemerellina** Lestage
s.g. **Austremerella** Riek (Aus)
s.g. **Ephemerellina** Lestage (Af, As)
s.g. **Lithogloea** Barnard (Af)
**Lestagella** Demoulin (Af)
**Manohyphella** Allen (Mad)
**Teloganella** Ulmer (As)
**Teloganodes** Eaton (As)
Subfamily incertae
†**Clephemera** Lin (Jurassic)
Leptohyphidae
Dicercomyzinae
**Dicercomyzon** Demoulin (Af)
Leptohyphinae
**Cotopaxi** Mayo (SA)
**Haplohyphes** Allen (SA)
**Leptohyphes** Eaton [= **Bruchella** Navás]
(NA, SA)
**Leptohyphodes** Ulmer (SA)
**Tricorythafer** Lestage [= **Caenopsis** Need-
ham; = **Needhamocaenis** Lestage]
(Af)
**Tricorythodes** Ulmer
s.g. **Homoleptohyphes** Allen & Murv-
osh (NA)
s.g. **Tricoryhyphes** Allen & Murvosh
(NA, SA)
s.g. **Tricorythodes** Ulmer (NA, SA)
**Tricorythopsis** Traver (SA)
Tricorythidae
Ephemerythinae
**Ephemerythus** Gillies
s.g. **Ephemerythus** Gillies (Af)
s.g. **Tricomerella** Demoulin (Af)
Machadorythinae
**Machadorythus** Demoulin (Af)

**Coryphorus** Peters (SA)

Tricorythinae

**Neurocaenis** Navás (Af, As, Mad)
**Tricorythus** Eaton
    s.g. **Tricorythurus** Lestage (Af)
    s.g. **Tricorythus** Eaton (Af)

Caenoidea
  Neoephemeridae

**Leucorhoenanthus** Ulmer [= **Caenomera**
    Demoulin] (E)
**Neoephemera** McDunnough
    s.g. **Neoephemera** McDunnough (As?,
    NA)
    s.g. **Oreianthus** Traver (NA)
**Neoephemeropsis** Ulmer (As)
**Potamanthellus** Lestage [= **Rhoenanth-
odes** Lestage] (As)

Caenidae

**Afrocaenis** Gillies (Af)
    **Afrocercus** Malzacher (Af)
**Amercaenis** Provansha & McCafferty (NA)
**Austrocaenis** Barnard (Af, Mad)
**Brachycercus** Curtis [= **Eurycaenis**
    Bengtsson] (As, E, NA, SA)
**Brasilocaenis** Puthz (SA)
**Caenis** Stephens [= **Ordella** Campion; =
    **Oxycypha** Burmeister] (Af, As, Aus, E,
    NA, SA)
**Caenoculis** Soldán (As)
**Caenodes** Ulmer (Af, As, E?)
**Caenomedea** Thew (Af, As)
**Caenopsella** Gillies (Af)
**Cercobrachys** Soldán (As, NA, SA)
**Clypeocaenis** Soldán (Af, As)
**Insulibrachys** Soldán (SA)
**Tasmanocoenis** Lestage [= **Pseudocaenis**
    Soldán] (Aus)

Baetiscoidea
  Baetiscidae

**Baetisca** Walsh
    s.g. **Baetisca** Walsh (NA)
    s.g. **Fascioculus** Pescador & Berner
    (NA)

Prosopistomatidae

**Prosopistoma** Latreille [= **Binoculus** Fourcroy; = **Chelysentomon** Joly & Joly] (Af, As, Aus, E, Mad)

INCERTAE SEDIS

†**Dyadentomum** Handlirsch (Permian)

†**Huizhougenia** Lin (Late Jurassic or Early Cretaceous)

†**Philolimnias** Hong (Eocene)

# ALPHABETICAL CATALOG
# OF FAMILY-GROUP NAMES

Acanthametropodinae Edmunds in Edmunds, Allen, & Peters, 1963:10.
Synonym Analetridinae Demoulin.

†Aenigmephemeridae Tshernova, 1968:23.

Ameletopsidae Edmunds, 1957:246 (as Ameletopsinae).

Ametropodidae Bengtsson, 1913:305 (as Ametropidae).

Analetridinae Demoulin, 1974:3. Junior synonym of Acanthametropodidae
Edmunds.

Anepeorinae Edmunds, 1962:10. Synonym Spinadinae Edmunds & Jensen.

†Aphelophlebodidae Pierce, 1945:3.

Arthropleinae Balthasar, 1937:204 (as Arthropleidae).

Asthenopodinae Edmunds & Traver, 1954:239.

Atalophlebiinae Peters, 1980:38.

Baetidae Leach, 1815:139 (as Baetida). Synonyms Callibaetidae Riek and
Cloeonidae Newman.

Baetiscidae Banks, 1900:246 (as Baetiscini).

Behningiidae Motaş & Bačesco, 1937:29.

Binoculidae Demoulin, 1954b:103. Junior synonym of Prosopistomatidae
Lameere.

†Bojophlebiidae Kukalová-Peck, 1985:934.

Brachycercidae Lestage, 1924c:62. Junior synonym of Caenidae Newman.

Caenidae Newman, 1853:187 (as Coenidae). Synonym Brachycercidae
Lestage.

Callibaetidae Riek, 1973:162 (as Callibaetinae).  Junior synonym of
    Baetidae Leach.

Campsurinae Traver in Needham, Traver, & Hsu, 1935:284.

Chiloporterinae Landa, 1973:156.

Chromarcyinae Demoulin, 1953(1940):8.    Synonym Pseudoligoneuriinae
    Ulmer.

Cloeonidae Newman 1853:187.  Junior synonym of Baetidae Leach.

Coloburiscinae Edmunds in Edmunds, Allen, & Peters, 1963:11.

Dicercomyzinae Edmunds & Traver, 1954:238.

Ecdyonuridae Ulmer, 1920a(1905):136.  Junior synonym of Heptageniidae
    Needham.

Ecdyuridae Jacobson & Bianchi, 1905:877.  Junior synonym of Heptageni-
    idae Needham.

†Epeoromimidae Tshernova, 1969:154.

Ephemerellidae Klapálek, 1909:13.

Ephemeridae Latreille, 1810:273 (as Ephemerinae).

†Ephemeropsidae Cockerell, 1924:136 (as Ephemeropsinae).

Ephemerythinae Gillies, 1960:35.

Ephoridae Traver in Needham, Traver, & Hsu, 1935:241 (as Ephoroninae).
    Junior synonym of Polymitarcyidae Banks.

†Eudoteridae Demoulin, 1954e:553.   Junior synonym of Misthodotidae
    Tillyard.

Euthyplociidae Lestage, 1921:213 (as Euthyplociinae).

Exeuthyplociinae Gillies, 1980:218.

Heptageniidae Needham in Needham & Betten, 1901:419 (as Heptageninae). Synonyms Ecdyonuridae Ulmer, Ecdyuridae Jacobson & Bianchi, and Rhithrogenidae Lestage.

†Hexagenitidae Lameere, 1917:74 (as Hexagenitinae). Synonyms Paedephemeridae Lameere and Stenodicranidae Demoulin.

Hyrtanellini Allen, 1980:88.

Ichthybotinae Demoulin, 1957:336 (as Ichthybotidae).

Isonychiinae Burks, 1953:108.

†Jarmilidae Demoulin, 1970b:7.

†Kukalovidae Demoulin, 1970b:6. Junior synonym of Protereismatidae Sellards.

Leptohyphidae Edmunds & Traver, 1954:238 (as Leptohyphinae).

Leptophlebiidae Banks, 1900:246 (as Leptophlebini).

†Litophlebiidae Hubbard & Riek, 1978:260. Replacement name for Xenophlebiidae Riek.
Litophlebiidae is now considered to belong to the Megasecoptera.

Machadorythinae Edmunds, Allen, & Peters, 1963:17.

Melanemerellinae Demoulin, 1955e:216.

†Mesephemeridae Lameere, 1917:47. Synonym Palingeniopsidae Martynov.

†Mesonetinae Tshernova, 1969:158 (as Mesonetidae).

†Mesoplectopteridae Demoulin, 1955j:345 (as Mesoplectopterinae).

Metretopodidae Traver in Needham, Traver, & Hsu, 1935:433 (as Metretopinae). Synonym Siphloplectidae Lestage.

†Misthodotidae Tillyard, 1932:260. Synonym Eudoteridae Demoulin.

Neoephemeridae Traver in Needham, Traver, & Hsu, 1935:288 (as Neoephemerinae).

†Oboriphlebiidae Hubbard & Kukalová-Peck, 1980:29.

Oligoneuriidae Ulmer, 1914:97.

Oniscigastridae Lameere, 1917:62 (as Oniscigastrina).

†Paedephemeridae Lameere, 1917:49. Junior synonym of Hexagenitidae Lameere.

Palingeniidae Albarda in Selys-Longchamps, 1888:147 (as Palingénines).

†Palingeniopsidae Martynov, 1938:35. Junior synonym of Mesephemeridae Lameere.

Pentageniinae McCafferty, 1972:51 (as Pentageniidae).

Polymitarcyidae Banks, 1900:246 (as Polymitarcini). Synonym Ephoridae Traver in Needham, Traver, & Hsu.

Potamanthidae Albarda in Selys-Longchamps, 1888:148 (as Potamanthines).

Prosopistomatidae Lameere, 1917:72 (as Prosopistomidae). Synonym Binoculidae Demoulin.

†Protereismatidae Sellards, 1907:345 (as Protereismephemeridae). Synonym Kukalovidae Demoulin.

Pseudironinae Edmunds & Traver, 1954:237.

Pseudoliogoneuriinae Ulmer, 1940:653. Junior synonym of Chromarcyinae Demoulin.

Rallidentinae Penniket, 1966:169.

Rhithrogenidae Lestage, 1917:266 (as Rhithrogeninae). Junior synonym of Heptageniidae Needham.

Siphlaenigmatidae Penniket, 1962:394.

Siphlonuridae Ulmer, 1920a(1888):131. Synonym Siphluridae Albarda.

Siphloplectidae Lestage, 1938a:180 (as Siphloplectonidae). Junior synonym of Metretopodidae Traver.

Siphluridae Albarda in Selys-Longchamps, 1888:150 (as Siphlurines). Junior synonym of Siphlonuridae Ulmer.

Spinadinae Edmunds & Jensen, 1974:495. Junior synonym of Anepeorinae Edmunds.

†Stenodicranidae Demoulin, 1954e:553. Junior synonym of Hexagenitidae Lameere.

†Syntonopteridae Handlirsch, 1911:299.

Teloganodinae Allen, 1965:263.

Timpanogae Allen, 1984:246.

†Torephemeridae Sinitshenkova, 1989:39.

Tricorythidae Lestage, 1942:15.

†Triplosobidae Handlirsch, 1906:312.

Vietnamellae Allen, 1984:246.

†Xenophlebiidae Riek, 1976:150. Junior synonym of Litophlebiidae Hubbard & Riek.
This family is now considered to belong to the Megasecoptera.

## ALPHABETICAL CATALOG OF GENUS-GROUP NAMES

### Genus **Acanthametropus** Tshernova
*Acanthametropus* Tshernova, 1948:1453. Type-species: *Acanthametropus nikolskyi* Tshernova, original designation. Siphlonuridae: Acanthametropodinae. Synonym *Metreturus* Burks.

### Genus **Acanthiops** Waltz & McCafferty
*Acanthiops* Waltz & McCafferty, 1987a:97. Type-species: *Centroptilum marlieri* Demoulin, original designation. Baetidae.

### Genus **Acanthomola** Whiting & Lehmkuhl
*Acanthomola* Whiting & Lehmkuhl, 1987b:410. Type-species: *Acanthomola pubescens* Whiting & Lehmkuhl, original designation. Heptageniidae: Heptageniinae.

### Genus **Acanthophlebia** Towns
*Acanthophlebia* Towns, 1983:28. Type-species: *Atalophlebia cruentata* Hudson, original designation. Leptophlebiidae: Atalophlebiinae.

### Subgenus **Acentrella** Bengtsson
*Acentrella* Bengtsson, 1912:110. Type-species: *Acentrella lapponica* Bengtsson, monotypy. Baetidae. Subgenus of *Baetis* Leach.
*Acentrella* has been treated as both a synonym and subgenus of *Baetis* Leach.

### Genus **Acerella** Allen
*Ephemerella* (*Acerella*) Allen, 1971:517. Type-species: *Ephemerella longicaudata* Uéno, original designation. Ephemerellidae: Ephemerellinae: Ephemerellini: Ephemerellae.
*Acerella* was originally proposed as a subgenus of *Ephemerella* Walsh.

### Genus **Acerpenna** Waltz & McCafferty
*Acerpenna* Waltz & McCafferty, 1987c:669. Type-species: *Baetis macdunnoughi* Ide, original designation. Baetidae.

### Genus **Adenophlebia** Eaton
*Adenophlebia* Eaton, 1881a:194. Type-species: *Ephemera dislocans* Walker, original designation. Leptophlebiidae: Atalophlebiinae. Synonym *Esbenophlebia* Lestage.

Genus **Adenophlebiodes** Ulmer

*Adenophlebiodes* Ulmer, 1924a:33. Type-species: *Adenophlebia ornata* Ulmer, original designation. Leptophlebiidae: Atalophlebiinae. Synonym *Euphlebia* Crass.

Two subgenera are recognized: *Adenophlebiodes* (s.s.) and *Hyalophlebia* Demoulin.

Genus **Aenigmephemera** Tshernova

*Aenigmephemera* Tshernova, 1968:23. Type-species: *Aenigmephemera demoulini* Tshernova, original designation. Aenigmephemeridae.

*Aenigmephemera* is known only from fossils.

Subgenus **Aethephemera** McCafferty & Edmunds

*Ephemera (Aethephemera)* McCafferty & Edmunds, 1973:306. Type-species: *Ephemera nadinae* McCafferty & Edmunds, original designation. Ephemeridae. Subgenus of *Ephemera* Linnaeus.

Genus *Afghanurus* Demoulin

*Afghanurus* Demoulin, 1964b:356. Type-species: *Afghanurus vicinus* Demoulin, original designation. Heptageniidae: Heptageniinae. Synonym of *Ecdyonurus* Eaton (Kluge, 1980:571).

Genus **Afrobaetodes** Demoulin

*Afrobaetodes* Demoulin, 1970d:52. Type-species: *Afrobaetodes berneri* Demoulin, original designation. Baetidae.

Genus **Afrocaenis** Gillies

*Afrocaenis* Gillies, 1982:15. Type-species: *Caenopsella major* Gillies, original designation. Caenidae.

Genus **Afrocercus** Malzacher

*Afrocercus* Malzacher, 1987:2. Type-species: *Afrocercus forcipatus* Malzacher, original designation. Caenidae.

Genus **Afromera** Demoulin

*Afromera* Demoulin, 1955i:292. Type-species: *Afromera congolana* Demoulin, original designation. Ephemeridae. Synonym *Dicrephemera* McCafferty & Edmunds.

*Afromera* has been treated as a synonym of *Ephemera* Linnaeus.

## Genus **Afronurus** Lestage

*Afronurus* Lestage, 1924b:349. Type-species: *Ecdyurus peringueyi* Esben-Petersen, original designation. Heptageniidae: Heptageniinae.

## Genus **Afroplocia** Lestage

*Afroplocia* Lestage, 1939:135. Type-species: *Exeuthyplocia sampsoni* Barnard, original designation. Euthyplociidae: Exeuthyplociinae.

## Subgenus *Akkarion* Flowers

*Nixe* (*Akkarion*) Flowers, 1980:303. Type-species: *Heptagenia simpliciodes* McDunnough, original designation. Leptophlebiidae: Atalophlebiinae. Synonym of *Ecdyonurus* Eaton (Tshernova et al., 1986:117).
*Akkarion* was originally described as a subgenus of *Nixe* Flowers.

## Genus **Albisca** Sinitshenkova

*Albisca* Sinitshenkova, 1989:39. Type-species: *Albisca tracheata* Sinitshenkova, original designation. Siphlonuridae: Acanthametropodidae.
*Albisca* is known only from fossils.

## Genus *Alloydia* Needham

*Alloydia* Needham, 1932:275. Type-species: *Alloydia cacautana* Needham, original designation. Oligoneuriidae: Oligoneuriinae. Synonym of *Lachlania* Hagen (Demoulin, 1952b:3).

## Genus **Ameletoides** Tillyard

*Ameletoides* Tillyard, 1933:5. Type-species: *Ameletoides lacusalbinae* Tillyard, original designation. Siphlonuridae: Siphlonurinae.

## Genus **Ameletopsis** Phillips

*Ameletopsis* Phillips, 1930a:324. Type-species: *Ameletus perscitus* Eaton, monotypy. Ameletopsidae: Ameletopsinae.

## Genus **Ameletus** Eaton

*Ameletus* Eaton, 1885:210. Type-species: *Ameletus subnotatus* Eaton, original designation. Siphlonuridae: Siphlonurinae. Synonyms *Chimura* Navás, *Metreletus* Demoulin, and *Paleoameletus* Lestage.

Genus **Amercaenis** Provansha & McCafferty
*Amercaenis* Provansha & McCafferty, 1985:2.   Type-species: *Caenis ridens* McDunnough, monotypy.   Caenidae.

Genus **Ametropus** Albarda
*Ametropus* Albarda, 1878:129.   Type-species: *Ametropus fragilis* Albarda, monotypy.   Ametropodidae.

Genus **Anagenesia** Eaton
*Anagenesia* Eaton, 1883:25.   Type-species: *Palingenia lata* Walker, original designation.   Palingeniidae: Palingeniinae.

Genus **Analetris** Edmunds
*Analetris* Edmunds in Edmunds & Koss, 1972:138.   Type-species: *Analetris eximia* Edmunds, original designation.   Siphlonuridae: Acanthametropodinae.

Genus *Andromina* Navás
*Andromina* Navás, 1912b:416. Type-species: *Andromina grisea* Navás, original designation.   Siphlonuridae: Siphlonurinae.   Synonym of *Siphlonurus* Eaton (Edmunds, 1960:24).

Genus **Anepeorus** McDunnough
*Anepeorus* McDunnough, 1925:190. Type-species: *Anepeorus rusticus* McDunnough, original designation.   Heptageniidae: Anepeorinae. Synonym *Spinadis* Edmunds & Jensen.

Genus **Aphelophlebodes** Pierce
*Aphelophlebodes* Pierce, 1945:3. Type-species: *Aphelophlebodes stocki* Pierce, original designation.   Aphelophlebodidae.
    *Aphelophlebodes* is known only from fossils.

Genus **Apobaetis** Day
*Apobaetis* Day, 1955:126.   Type-species: *Apobaetis indeprensus* Day, original designation.   Baetidae.

Genus **Aprionyx** Barnard
*Aprionyx* Barnard, 1940:616.   Type-species: *Atalophlebia tabularis* Eaton, original designation.   Leptophlebiidae: Atalophlebiinae.
    *Aprionyx* was originally proposed by Barnard in 1932. The name did not become available until the type-species was designated in 1940.

## Genus **Arachnocolus** Towns & Peters

*Arachnocolus* Towns & Peters, 1979b:444. Type-species: *Arachnocolus phillipsi* Towns & Peters, original designation. Leptophlebiidae: Atalophlebiinae.

## Genus **Archeobehningia** Tshernova

*Archeobehningia* Tshernova, 1977:94. Type-species: *Archeobehningia edmundsi* Tshernova, original designation. Behningiidae. *Archeobehningia* is known only from fossils.

## Genus **Archethraulodes** Pescador & Peters

*Archethraulodes* Pescador & Peters, 1982:1. Type-species: *Archethraulodes spatulus* Pescador & Peters, original designation. Leptophlebiidae: Atalophlebiinae.

## Genus **Arthroplea** Bengtsson

*Arthroplea* Bengtsson, 1908:239. Type-species: *Arthroplea congener* Bengtsson, monotypy. Heptageniidae: Arthropleinae. Synonyms *Haplogenia* Blair and *Remipalpus* Bengtsson.

## Genus *Asiatella* Tshernova

*Asiatella* Tshernova, 1972:611. Type-species: *Ephemerella nigra* Uéno, original designation. Ephemerellidae: Ephemerellinae: Ephemerellini: Ephemerellae. Synonym of *Cincticostella* Allen (Tshernova, 1972:614).

## Genus **Asionurus** Braasch & Soldán

*Asionurus* Braasch & Soldán, 1986b:155. Type-species: *Asionurus primus* Braasch & Soldán, original designation. Heptageniidae: Heptageniinae

## Genus **Askola** Peters

*Askola* Peters, 1969:253. Type-species: *Askola froehlichi* Peters, original designation. Leptophlebiidae: Atalophlebiinae.

## Genus **Asthenopodes** Ulmer

*Asthenopodes* Ulmer, 1924a:26. Type-species: *Asthenopous albicans* Pictet nec Percheron [= *Asthenopodes picteti* Hubbard], original designation. Polymitarcyidae: Asthenopodinae. Synonym of *Asthenopus* Eaton (Hubbard & Dominguez, 1988:107).

## Genus **Asthenopodichnium** Thenius

*Asthenopodichnium* Thenius, 1979:185. Type-species: *Asthenopodichnium xylobiontum* Thenius, original designation. Polymitarcyidae: Asthenopodinae.

Asthenopodichnium is known only from fossils.

## Genus **Asthenopus** Eaton

*Asthenopus* Eaton, 1871:59. Type-species: *Palingenia curta* Hagen, original designation. Polymitarcyidae: Asthenopodinae. Synonym *Asthenopodes* Ulmer.

## Genus **Atalomicria** Harker

*Atalomicria* Harker, 1954:252. Type-species: *Atalophlebia uncinata* Ulmer, original designation. Leptophlebiidae: Atalophlebiinae.

## Genus *Atalonella* Needham & Murphy

*Atalonella* Needham & Murphy, 1924:35. Type-species: *Atalonella ophis* Needham & Murphy, subsequent designation by Peters & Edmunds, 1972:1411. Leptophlebiidae: Atalophlebiinae. Synonym of *Nousia* Navás (Pescador & Peters, 1985:93).

## Genus **Atalophlebia** Eaton

*Atalophlebia* Eaton, 1881a:193. Type-species: *Ephemera australis* Walker, original designation. Leptophlebiidae: Atalophlebiinae.

## Genus **Atalophlebioides** Phillips

*Deleatidium* (*Atalophlebioides*) Phillips, 1930b:336. Type-species: *Deleatidium cromwelli* Phillips, subsequent designation by Peters & Edmunds, 1964:238. Leptophlebiidae: Atalophlebiinae.

Atalophlebioides was originally proposed as a subgenus of Deleatidum Eaton.

## Genus **Atopophlebia** Flowers

*Atopophlebia* Flowers, 1980:162. Type-species: *Atopophlebia fortunensis* Flowers, original designation. Leptophlebiidae: Atalophlebiinae.

## Genus **Atopopus** Eaton

*Atopopus* Eaton, 1881b:22. Type-species: *Atopopus tarsalis* Eaton, original designation. Heptageniidae: Heptageniinae.

## Subgenus **Atroari** Savage & Peters

*Miroculis (Atroari)* Savage & Peters, 1983:549. Type-species: *Miroculis duckensis* Savage & Peters, original designation. Leptophlebiidae: Atalophlebiinae. Subgenus of *Miroculis* Edmunds.

## Genus **Attenella** Edmunds

*Ephemerella (Attenella)* Edmunds, 1971:152. Type-species: *Ephemerella attenuata* McDunnough, objective synonymy. Ephemerellidae: Ephemerellinae: Ephemerellini: Timpanogae. Replacement name for *Attenuatella* Edmunds.

## Genus *Attenuatella* Edmunds

*Ephemerella (Attenuatella)* Edmunds, 1959:546. Type-species: *Ephemerella attenuata* McDunnough, original designation. Preoccupied. Ephemerellidae: Ephemerellinae: Ephemerellini: Timpanogae. Objective synonym of *Attenella* Edmunds.

*Attenuatella* was originally proposed as a subgenus of *Ephemerella* Walsh.

## Subgenus **Australonousia** Campbell & Suter

*Nousia (Australonousia)* Campbell & Suter, 1988:270. Type-species: *Atalophlebia fuscula* Tillyard, original designation. Leptophlebiidae: Atalophlebiinae. Subgenus of *Nousia* Navás.

## Subgenus **Australphlebia** Peters

*Borinquena (Australphlebia)* Peters, 1971:27. Type-species: *Borinquena traverae* Peters, original designation. Leptophlebiidae: Atalophlebiinae. Subgenus of *Borinquena* Traver.

## Genus **Australurus** Jell & Duncan

*Australurus* Jell & Duncan, 1986:120. Type-species: *Australurus plexus* Jell & Duncan, original designation. Siphlonuridae.

*Australurus* is known only from fossils.

## Subgenus **Austremerella** Riek

*Austremerella* Riek, 1963:50. Type-species: *Austremerella picta* Riek, original designation. Ephemerellidae: Teloganodinae. Subgenus of *Ephemerellina* Lestage.

## Genus **Austrocaenis** Barnard

*Austrocaenis* Barnard, 1932:227. Type-species: *Austrocaenis capensis* Barnard, monotypy. Caenidae.

## Genus **Austroclima** Towns & Peters
*Austroclima* Towns & Peters, 1979a:213.   Type-species: *Deleatidium sepia* Phillips, original designation. Leptophlebiidae: Atalophlebiinae.

## Genus *Austrocloeon* Barnard
*Austrocloeon* Barnard, 1940:616.   Type-species: *Cloeon africanum* Esben-Petersen, original designation. Baetidae.  Synonym of *Cloeon* Leach (Demoulin, 1970d:53).
*Austrocloeon* was originally proposed by Barnard in 1932. The name did not become available until the type-species was designated in 1940.

## Genus **Austrophlebioides** Campbell & Suter
*Austrophlebioides* Campbell & Suter, 1988:260.   Type-species: *Deleatidium pussilum* Harker, original designation. Leptophlebiidae: Atalophlebiinae.

## Subgenus **Baetiella** Uéno
*Baetiella* Uéno, 1931:220.   Type-species: *Acentrella japonica* Imanishi, original designation.   Baetidae.   Synonym *Neobaetiella* Müller-Liebenau.   Subgenus of *Baetis* Leach.
*Baetiella* has been considered a synonym of *Pseudocloeon* Klapálek.

## Genus **Baetis** Leach
*Baetis* Leach, 1815:137.   Type-species: *Ephemera fuscata* Linnaeus, subsequent designation by International Commission on Zoological Nomenclature, 1966:209.  Baetidae.  Synonym *Brachyphlebia* Westwood.
Six subgenera of *Baetis* are recognized: *Baetis* (s.s.) Leach, *Acentrella* Bengtsson, *Baetiella* Uéno, *Labiobaetis* Novikova & Kluge, *Nigrobaetis* Kazlauskas, and *Takobia* Novikova & Kulge.
Three subgenera of *Baetis* (in addition to the valid subgenera listed above) were proposed by Kazlauskas (1972): *Montobaetis* Kazlauskas, *Rhodobaetis* Kazlauskas, and *Vernobaetis* Kazlauskas.  *Holobaetis* Sukatskene was also proposed as a subgenus of *Baetis* (Sukatskene, 1962).  None of these subgenera are available because no type-species was designated for any of them.

## Genus **Baetisca** Walsh
*Baetisca* Walsh, 1863b:378.   Type-species: *Baetis obesa* Say, monotypy. Baetiscidae.

## Genus **Baetodes** Needham & Murphy
*Baetodes* Needham & Murphy, 1924:55.   Type-species: *Baetodes serratus* Needham & Murphy, original designation. Baetidae.

## Genus **Baetopus** Keffermüller

*Baetopus* Keffermüller, 1960:425. Type-species: *Baetopus wartensis* Keffermüller, monotypy. Baetidae.
Two subgenera are recognized: *Baetopus* (s.s.) and *Raptobaetopus* Müller-Liebenau.

## Genus **Baltameletus** Demoulin

*Baltameletus* Demoulin, 1968b:238. Type-species: *Baltameletus oligocaenicus* Demoulin, original designation. Siphlonuridae: Siphlonurinae.
*Baltameletus* is known only from fossils.

## Genus **Balticophlebia** Demoulin

*Balticophlebia* Demoulin, 1968b:237. Type-species: *Balticophlebia hennigi* Demoulin, original designation. Ameletopsidae: Ameletopsinae.
*Balticophlebia* is known only from fossils.

## Genus *Bantiska* Sellards

*Bantiska* Sellards, 1907:349. Type-species: *Bantiska elongata* Sellards, original designation. Protereismatidae. Synonym of *Protereisma* Sellards (Tillyard, 1932:244).

## Genus **Barbaetis** Waltz & McCafferty

*Barbaetis* Waltz & McCafferty in Waltz, McCafferty, & Kennedy, 1985:161. Type-species: *Barbaetis benfieldi* Kennedy in Waltz, McCafferty, & Kennedy, original designation. Baetidae.

## Genus **Behningia** Lestage

*Behningia* Lestage, 1930:436. Type-species: *Behningia ulmeri* Lestage, original designation. Behningiidae.

## Genus **Belovius** Tshernova

*Belovius* Tshernova, 1981:326. Type-species: *Epeorus latifolium* Uéno, original designation. Heptageniidae: Heptageniinae.

## Genus **Bernerius** Waltz & McCafferty

*Bernerius* Waltz & McCafferty, 1987b:179. Type-species: *Bernerius incus* Waltz & McCafferty, original designation. Baetidae.

## Genus **Bibulmena** Dean
*Bibulmena* Dean, 1987:96. Type-species: *Bibulmena kadjina* Dean, original designation. Leptophlebiidae: Atalophlebiinae.

## Genus **Binoculus** Geoffroy
*Binoculus* Geoffroy, 1762:658. Supressed by International Commission on Zoological Nomenclature, 1954:211. Prosopistomatidae. Synonym of *Prosopistoma* Latreille.

## Genus **Blanchardia** Brongniart
*Blanchardia* Brongniart, 1893[1894]:325. Type-species: *Blanchardia pulchella* Brongniart, monotypy. Preoccupied. Triplosobidae. Objective synonym of *Triplosoba* Handlirsch.

## Genus **Blasturophlebia** Demoulin
*Blasturophlebia* Demoulin, 1968b:268. Type-species: *Blasturophlebia hirsuta* Demoulin, original designation. Leptophlebiidae: Atalophlebiinae.
*Blasturophlebia* is known only from fossils.

## Genus **Blasturus** Eaton
*Blasturus* Eaton, 1881a:193. Type-species: *Ephemera cupida* Say, original designation. Leptophlebiidae: Leptophlebiinae. Synonym of *Leptophlebia* Westwood (Ide, 1935:124).

## Genus **Bleptus** Eaton
*Bleptus* Eaton, 1885:243. Type-species: *Bleptus fasciatus* Eaton, original designation. Heptageniidae: Heptageniinae.

## Genus **Bojophlebia** Kukalová-Peck
*Bojophlebia* Kukalová-Peck, 1985:936. Type species: *Bojophlebia prokopi* Kukalová-Peck, original designation. Bojophlebiidae.
*Bojophlebia* is known only from fossils.

## Genus **Borinquena** Traver
*Borinquena* Traver, 1938:16. Type-species: *Borinquena carmencita* Traver, original designation. Leptophlebiidae: Atalophlebiinae.
Two subgenera are recognized: *Borinquena* (s.s.) and *Australphlebia* Peters.

## Subgenus **Borisonychia** McCafferty
*Isonychia (Borisonychia)* McCafferty, 1989:78. Type-species: *Isonychia diversa* Traver, original designation. Oligoneuriidae: Isonychiinae. Subgenus of *Isonychia* Eaton.

## Genus **Brachycercus** Curtis
*Brachycercus* Curtis, 1834:122. Type-species: *Brachycercus harrisellus* Curtis, subsequent designation by Lestage, 1924b:61. Caenidae. Synonym *Eurycaenis* Bengtsson.

## Genus *Brachyphlebia* Westwood
*Brachyphlebia* Westwood, 1840:25. Type-species: *Ephemera fuscata* Linnaeus [misidentified as *Ephemera bioculata*], monotypy. Baetidae. Objective synonym of *Baetis* Leach.

## Genus **Brasilocaenis** Puthz
*Brasilocaenis* Puthz, 1975:411. Type-species: *Brasilocaenis irmleri* Puthz, original designation. Caenidae.

## Genus **Brevitibia** Demoulin
*Brevitibia* Demoulin, 1968b:245. Type-species: *Brevitibia intricans* Demoulin, original designation. Ametropodidae.
*Brevitibia* is known only from fossils.

## Genus *Bruchella* Navás
*Bruchella* Navás, 1920:56. Type-species: *Bruchella nigra* Navás, original designation. Leptohyphidae: Leptohyphinae. Synonym of *Leptohyphes* Eaton (Edmunds, Allen, & Peters, 1963:17).

## Genus **Bungona** Harker
*Bungona* Harker, 1957:73. Type-species: *Bungona narilla* Harker, original designation. Baetidae.

## Genus **Caenis** Stephens
*Caenis* Stephens, 1835:60. Type-species: *Caenis macrura* Stephens, subsequent designation by Westwood, 1840:synopsis p. 47. Caenidae. Synonyms *Ordella* Campion and *Oxycypha* Burmeister.

Genus **Caenoculis** Soldán
*Caenoculis* Soldán, 1986:347. Type-species: *Caenoculis bishopi* Soldán,
original designation. Caenidae.

Genus **Caenodes** Ulmer
*Caenodes* Ulmer, 1924b:7. Type-species: *Caenodes ulmeri* Kimmins [=
*Caenis cibaria* sensu Ulmer], subsequent designation by Kimmins,
1949:831. Caenidae.

Genus **Caenomedea** Thew
*Caenomedea* Thew, 1960:199. Type-species: *Caenis kivuensis* Demoulin, original designation. Caenidae.

Subgenus *Caenomera* Demoulin
*Neoephemera* (*Caenomera*) Demoulin, 1961:66. Type-species: *Caenis
maxima* Joly, original designation. Neoephemeridae. Synonym of
*Leucorhoenanthus* Lestage (Demoulin, 1962:369).
    ' *Caenomera* was originally proposed as a subgenus of *Neoephemera* McDunnough.

Genus **Caenopsella** Gillies
*Caenopsella* Gillies, 1977:451. Type-species: *Caenopsella meridies*
Gillies, original designation. Caenidae.

Genus *Caenopsis* Needham
*Caenopsis* Needham, 1920:39. Type-species: *Caenopsis fugitans*
Needham, monotypy. Preoccupied. Leptohyphidae: Leptohyphinae.
Objective synonym of *Tricorythafer* Lestage.
    Lestage proposed two separate replacement names for the preoccupied *Caenopsis*:
*Tricorythafer* (1942:4) and *Needhamocoenis* (1945:85).

Genus **Calliarcys** Eaton
*Calliarcys* Eaton, 1881a:196. Type-species: *Calliarcys humilis* Eaton,
original designation. Leptophlebiidae: Leptophlebiinae.

Genus **Callibaetis** Eaton
*Callibaetis* Eaton, 1881a:196. Type-species: *Baetis pictus* Eaton,
original designation. Baetidae. Synonym *Neobaetis* Navás.

Genus **Camelobaetidius** Demoulin
*Camelobaetidius* Demoulin, 1966a:9. Type-species: *Camelobaetidius
leentvaari* Demoulin, original designation. Baetidae.

## Genus **Campsurus** Eaton
*Campsurus* Eaton, 1868a:83.    Type-species: *Palingenia latipennis*
Walker, original designation.    Polymitarcyidae: Campsurinae.

## Genus **Campylocia** Needham & Murphy
*Campylocia* Needham & Murphy, 1924:25.    Type-species: *Euthyplocia
anceps* Eaton, original designation.    Euthyplociidae: Euthyplociinae.
Synonym *Longinella* Gros & Lestage.

## Genus **Careospina** Peters
*Careospina* Peters, 1971:11.    Type-species: *Careospina hespera* Peters
& Alayo, original designation.    Leptophlebiidae: Atalophlebiinae.

## Genus **Castanophlebia** Barnard
*Castanophlebia* Barnard, 1932:244.    Type-species: *Castanophlebia
calida* Barnard, monotypy.    Leptophlebiidae: Atalophlebiinae.

## Genus **Caudatella** Edmunds
*Ephemerella* (*Caudatella*) Edmunds, 1959:546. Type-species: *Ephemerella heterocaudata* McDunnough, original designation.    Ephemerellidae: Ephemerellinae: Ephemerellini: Ephemerellae.
   *Caudatella* was originally proposed as a subgenus of *Ephemerella* Walsh.

## Genus **Caurinella** Allen
*Caurinella* Allen, 1984:245. Type-species: *Caurinella idahoensis* Allen,
original designation.    Ephemerellidae: Ephemerellinae: Ephemerellini.

## Genus **Celiphlebia** Peters & Peters
*Celiphlebia* Peters & Peters, 1980:61.    Type-species: *Celiphlebia
caledonae* Peters & Peters, original designation.    Leptophlebiidae:
Atalophlebiinae.

## Genus *Centroptella* Braasch & Soldán
*Centroptella* Braasch & Soldán, 1980:123. Type-species: *Centroptella
longisetosa* Braasch & Soldán, monotypy.    Baetidae.    Synonym of
*Cloeodes* Traver (Waltz & McCafferty, 1987b:177).

Genus **Centroptiloides** Lestage
*Centroptiloides* Lestage, 1918:107. Type-species: *Centroptilum bifasciatum* Esben-Petersen, original designation. Baetidae. Synonym *Haplobaetis* Navás.

Genus **Centroptilum** Eaton
*Centroptilum* Eaton, 1869:132. Type-species: *Ephemera luteola* Müller, original designation. Baetidae.

Genus **Cercobrachys** Soldán
*Cercobrachys* Soldán, 1986:336. Type-species: *Cercobrachys etowah* Soldán, original designation. Caenidae.

Genus **Chankagenesia** Buldovski
*Chankagenesia* Buldovski, 1935:831. Type-species: *Chankagenesia natans* Buldovski, original designation. Palingeniidae: Palingeniinae.

Genus **Chaquihua** Demoulin
*Chaquihua* Demoulin, 1955b:11. Type-species: *Chaquihua penai*, original designation. Ameletopsidae: Ameletopsinae.

Genus **Cheirogenesia** Demoulin
*Cheirogenesia* Demoulin, 1952a:10. Type-species: *Anagenesia decaryi* Navás, original designation. Palingeniidae: Palingeniinae. Synonym *Fontanica* McCafferty.

Genus *Chelysentomon* Joly & Joly
*Chelysentomon* Joly & Joly, 1872:438. Type-species: *Prosopistoma varigatum* Latreille, objective synonymy. Unnecessary replacement name for *Prosopistoma* Latreille. Prosopistomatidae. Objective synonym of *Prosopistoma* Latreille.

Genus **Chiloporter** Lestage
*Chiloporter* Lestage, 1931b:50. Type-species: *Chiloporter eatoni* Lestage, original designation. Ameletopsidae: Chiloporterinae.

Genus *Chimura* Navás
*Chimura* Navás, 1915:149. Type-species: *Chimura aetherea* Navás, original designation. Siphlonuridae: Siphlonurinae. Synonym of *Ameletus* Eaton (Edmunds, 1960:24).

## Genus *Chirotonetes* Eaton

*Chirotonetes* Eaton, 1881b:21. Type-species: *Isonychia manca* Eaton, objective synonymy. Unnecessary replacement name for *Isonychia* Eaton. Oligoneuriidae: Isonychiinae. Objective synonym of *Isonychia* Eaton.

## Genus *Chitonophora* Bengtsson

*Chitonophora* Bengtsson, 1908:243. Type-species: *Chitonophora aurivillii* Bengtsson, monotypy. Ephemerellidae: Ephemerellinae: Ephemerellini: Ephemerellae. Synonym of *Ephemerella* Walsh (Allen & Edmunds, 1962:246).
*Chitonophora* has been treated as a subgenus of *Ephemerella* Walsh.

## Genus **Chiusanophlebia** Uéno

*Chiusanophlebia* Uéno, 1969:230. Type-species: *Chiusanophlebia asahinai* Uéno, original designation. Leptophlebiidae: Atalophlebiinae.

## Genus **Chopralla** Waltz & McCafferty

*Chopralla* Waltz & McCafferty, 1987b:182. Type-species: *Centroptella ceylonensis* Müller-Liebenau, original designation. Baetidae.

## Genus **Choroterpes** Eaton

*Choroterpes* Eaton, 1881a:194. Type-species: *Choroterpes lusitanica* Eaton, original designation. Leptophlebiidae: Atalophlebiinae.
Three subgenera are recognized: *Choroterpes* (s.s.), *Euthraulus* Barnard, and *Neochoroterpes* Allen.

## Genus **Choroterpides** Ulmer

*Choroterpides* Ulmer, 1939:494. Type-species: *Thraulus exiguus* Eaton, original designation. Leptophlebiidae: Atalophlebiinae.

## Genus **Chromarcys** Navás

*Chromarcys* Navás, 1932:927. Type-species: *Chromarcys magnifica* Navás, original designation. Oligoneuriidae: Chromarcyinae. Synonym *Pseudoligoneuria* Ulmer.

## Genus **Cincticostella** Allen

*Ephemerella* (*Cincticostella*) Allen, 1971:513. Type-species: *Ephemerella nigra* Uéno, original designation. Ephemerellidae: Ephemerellinae: Ephemerellini: Ephemerellae. Synonym *Asiatella* Tshernova.

Two subgenera are recognized: *Cincticostella* Allen and *Rhionella* Allen. *Cincticostella* was originally proposed as a subgenus of *Ephemerella* Walsh.

## Genus **Cinygma** Eaton
*Cinygma* Eaton, 1885:247. Type-species: *Cinygma integrum* Eaton, original designation. Heptageniidae: Heptageniinae.

## Genus *Cinygmina* Kimmins
*Cinygmina* Kimmins, 1937:435. Type-species: *Cinygmina assamensis* Kimmins, original designation. Heptageniidae: Heptageniinae. Synonym of *Ecdyonurus* Eaton (Tshernova et al., 1986:114).

## Genus **Cinygmoides** Matsumura
*Cinygmoides* Matsumura, 1931:1474. Type-species: *Cinygmoides hekachii* Matsumura, monotypy. Heptageniidae: Heptageniinae.

## Genus **Cinygmula** McDunnough
*Cinygmula* McDunnough, 1933:75. Type-species: *Ecdyurus ramaleyi* Dodds, original designation. Heptageniidae: Heptageniinae.

## Genus **Clephemera** Lin
*Clephemera* Lin, 1986:27. Type-species: *Clephemera clava* Lin, original designation. Ephemerellidae.

*Clephemera* is known only from fossils.

## Genus *Cloe* Burmeister
*Cloe* Burmeister, 1839:797. Type-species:*Ephemera diptera* Linnaeus, objective synonymy. Unnecessary replacement name for *Cloeon* Leach. Baetidae. Objective synonym of *Cloeon* Leach.

## Genus **Cloeodes** Traver
*Cloeodes* Traver, 1938:32. Type-species: *Cloeodes maculatus* Traver, original designation. Baetidae. Synonym *Centroptella* Braasch and Soldán.

Two subgenera are recognized: *Cloeodes* (s.s.) and *Notobaetis* Morihara & Edmunds.

## Genus **Cloeon** Leach
*Cloeon* Leach, 1815:137. Type-species:*Ephemera diptera* Linnaeus [as *Cloeon pallida* Leach], monotypy. Baetidae. Synonyms *Austrocloeon* Barnard, *Cloe* Burmeister, *Cloeopsis* Eaton, and *Neocloeon* Traver.

## Genus *Cloeopsis* Eaton
*Cloeopsis* Eaton, 1866:146. Type-species: *Ephemera diptera* Linnaeus, monotypy. Baetidae. Objective synonym of *Cloeon* Leach.

## Genus *Cloeoptilum* Kazlauskas
*Cloeoptilum* Kazlauskas, 1972:338. Not available, Art. 13b, no type designated. Baetidae. Synonym of *Pseudocentroptilum* Bogoesco (Keffermüller & Sowa, 1984:311).

## Genus **Clypeocaenis** Soldán
*Clypeocaenis* Soldán, 1978:119. Type-species: *Clypeocaenis bisetosa* Soldán, original designation. Caenidae.

## Genus **Coloburiscoides** Lestage
*Coloburiscoides* Lestage, 1935:356. Type-species: *Coloburiscus giganteus* Tillyard, original designation. Oligoneuriidae: Coloburiscinae.

## Genus **Coloburiscus** Eaton
*Coloburiscus* Eaton, 1888:349. Type-species: *Palingenia humeralis* Walker, objective synonymy. Oligoneuriidae: Coloburiscinae. Replacement name for *Coloburus* Eaton.

## Genus *Coloburus* Eaton
*Coloburus* Eaton, 1868a:89. Type-species: *Palingenia humeralis* Walker, original designation. Preoccupied. Oligoneuriidae: Coloburiscinae. Objective synonym of *Coloburiscus* Eaton.

## Genus **Compsoneuria** Eaton
*Compsoneuria* Eaton, 1881b:23. Type-species: *Compsoneuria spectabilis* Eaton, original designation. Heptageniidae: Heptageniinae. Synonyms *Compsoneuriella* Ulmer, *Cosmetogenia* Eaton, and *Notonurus* Crass.

## Genus *Compsoneuriella* Ulmer
*Compsoneuriella* Ulmer, 1939:563. Type-species: *Compsoneuriella thienemanni* Ulmer, original designation. Heptageniidae: Heptageniinae. Synonym of *Compsoneuria* Eaton (Braasch & Soldán, 1986a:46).

## Genus **Coryphorus** Peters
*Coryphorus* Peters, 1981:209.    Type-species: *Coryphorus aquilus*
Peters, original designation. Tricorythidae: Machadorythinae.

## Genus *Cosmetogenia* Eaton
*Cosmetogenia* Eaton, 1883:19, pl. 23.    Type-species: *Compsoneuria*
*spectabilis* Eaton, monotypy.    Heptageniidae: Heptageniinae.
Objective synonym of *Compsoneuria* Eaton.

## Genus **Cotopaxi** Mayo
*Cotopaxi* Mayo, 1968:301.    Type-species: *Cotopaxi macuchae* Mayo,
original designation. Leptohyphidae: Leptohyphinae.

## Genus **Coula** Peters & Peters
*Coula* Peters & Peters, 1980:73.    Type-species: *Coula fasciata* Peters
& Peters, original designation. Leptophlebiidae: Atalophlebiinae.

## Genus **Cretoneta** Tshernova
*Cretoneta* Tshernova, 1971:614.    Type-species: *Cretoneta zherichini*
Tshernova, original designation. Leptophlebiidae: Mesonetinae.
*Cretoneta* is known only from fossils.

## Genus **Crinitella** Allen & Edmunds
*Ephemerella* (*Crinitella*) Allen & Edmunds, 1963:17.    Type-species:
*Ephemerella coheri* Allen & Edmunds, original designation. Ephemer-
ellidae: Ephemerellinae: Ephemerellini: Ephemerellae.
*Crinitella* was originally proposed as a subgenus of *Ephemerella* Walsh.

## Genus **Cronicus** Eaton
*Cronicus* Eaton, 1871:133.    Type-species: *Baetis anomala* Pictet,
monotypy. Oligoneuriidae: Coloburiscinae.
*Cronicus* is known only from fossils.

## Genus **Cryophlebia** Towns & Peters
*Cryophlebia* Towns & Peters, 1979a:230.    Type-species: *Atalophleb-*
*ioides aucklandensis* Peters, original designation. Leptophlebiidae:
Atalophlebiinae.

## Genus **Cryptopenella** Gillies
*Cryptopenella* Gillies, 1951:125.    Type-species: *Cryptopenella facialis*
Gillies, original designation. Leptophlebiidae: Atalophlebiinae.

## Subgenus **Dacnogenia** Kluge
*Heptagenia (Dacnogenia)* Kluge, 1987:303. Type-species: *Heptagenia coerulans* Rostock, monotypy). Heptageniidae: Heptageniinae. Subgenus of *Heptagenia* Walsh.

## Genus **Dactylobaetis** Traver & Edmunds
*Dactylobaetis* Traver & Edmunds, 1968:629. Type-species: *Dactylobaetis warreni* Traver & Edmunds, original designation. Baetidae.

## Genus **Dactylophlebia** Pescador & Peters
*Dactylophlebia* Pescador & Peters, 1980b:332. Type-species: *Dactylophlebia carnulenta* Pescador & Peters, original designation. Leptophlebiidae: Atalophlebiinae.

## Genus **Dannella** Edmunds
*Ephemerella (Dannella)* Edmunds, 1959:546. Type-species: *Ephemerella simplex* McDunnough, original designation. Ephemerellidae: Ephemerellinae: Ephemerellini: Timpanogae.
Two subgenera are recognized: *Dannella* (s.s.) and *Dentatella* Allen. *Dannella* was originally proposed as a subgenus of *Ephemerella* Walsh.

## Genus **Deleatidium** Eaton
*Deleatidium* Eaton, 1899:288. Type-species: *Deleatidium lillii* Eaton, monotypy. Leptophlebiidae: Atalophlebiinae.

## Genus **Demoulinellus** Pescador & Peters
*Demoulinellus* Pescador & Peters, 1982:10. Type-species: *Demoulinellus coloratus* Pescador & Peters, original designation. Leptophlebiidae: Atalophlebiinae.

## Genus **Denina** McCafferty
*Denina* McCafferty, 1987:472. Type-species: *Denina dubiloca* McCafferty, original designation. Ephemeridae.
*Denina* is known only from fossils.

## Subgenus **Dentatella** Allen
*Dannella (Dentatella)* Allen, 1980:88. Type-species: *Ephemerella bartoni* Allen, original designation. Ephemerellidae: Ephemerellinae: Ephemerellini: Timpanogae. Subgenus of *Dannella* Edmunds.

## Genus **Dicercomyzon** Demoulin

*Dicercomyzon* Demoulin, 1954a:1. Type-species: *Dicercomyzon femorale* Demoulin, original designation. Leptohyphidae: Dicercomyzinae.

## Genus *Dicrephemera* McCafferty & Edmunds

*Ephemera (Dicrephemera)* McCafferty & Edmunds, 1973:302. Type-species: *Ephemera siamensis* Uéno, original designation. Ephemeridae. Synonym of *Afromera* Demoulin (McCafferty & Gillies, 1979: 170).
*Dicrephemera* was originally proposed as a subgenus of *Ephemera* Linnaeus.

## Genus *Dictyosiphlon* Lestage

*Dictyosiphlon* Lestage, 1931b:47. Type-species: *Heptagenia molinai* Navás, monotypy. Oligoneuriidae: Coloburiscinae. Synonym of *Murphyella* Lestage (Hubbard, 1985:12).

## Genus **Diphetor** Waltz & McCafferty

*Diphetor* Waltz & McCafferty, 1987c:669. Type-species: *Baetis hageni* Eaton, original designation. Baetidae.

## Genus *Dipteromimcdes* Matsumura

*Dipteromimodes* Matsumura, 1931:1474. Type-species: *Dipteromimodes suzukii* Matsumura, monotypy. Siphlonuridae: Siphlonurinae. Synonym of *Dipteromimus* McLachlan (Uéno, 1931:216).

## Genus **Dipteromimus** McLachlan

*Dipteromimus* McLachlan, 1875:170. Type-species: *Dipteromimus tipuliformis* McLachlan, monotypy. Siphlonuridae: Siphlonurinae. Synonym *Dipteromimodes* Matsumura.

## Genus **Dipterophlebiodes** Demoulin

*Dipterophlebiodes* Demoulin, 1954c:129. Type-species: *Dipterophlebiodes sarawacensis* Demoulin, original designation. Leptophlebiidae: Leptophlebiinae.

## Genus **Dolania** Edmunds & Traver

*Dolania* Edmunds & Traver, 1959:46. Type-species: *Dolania americana* Edmunds & Traver, original designation. Behningiidae.

## Genus *Doter* Sellards

*Doter* Sellards, 1907:355. Type-species: *Doter minor* Sellards, original designation.
Originally proposed as a mayfly, *Doter* is now considered as a Palaeoptera, *incerti ordinis*, by Carpenter (1980)

## Genus *Dromeus* Sellards

*Dromeus* Sellards, 1907:351. Type-species: *Dromeus obtusus* Sellards, original designation. Preoccupied. Misthodotidae. Objective synonym of *Misthodotes* Sellards.

## Genus **Drunella** Needham

*Drunella* Needham, 1905:42. Type-species: *Ephemerella grandis* Eaton, monotypy. Ephemerellidae: Ephemerellinae: Ephemerellini: Ephemerellae.
Five subgenera are recognized: *Drunella* (s.s.), *Eatonella* Needham, *Myllonella* Allen, *Tribrochella* Allen, and *Unirachella* Allen. *Drunella* has been treated as a subgenus of *Ephemerella* Walsh.

## Genus **Dulcimanna** Jell & Duncan

*Dulcimanna* Jell & Duncan, 1986:124. Type-species: *Dulcimanna sculptor* Jell & Duncan, original designation. Siphlonuridae.
*Dulcimanna* is known only from fossils.

## Genus **Dyadentomum** Handlirsch

*Dyadentomum* Handlirsch, 1904:7. Type-species: *Dyadentomum permense* Handlirsch, original designation. Family incertus.
*Dyadentomum* is known only from fossils.

## Subgenus **Eatonella** Needham

*Ephemerella (Eatonella)* Needham, 1927:108. Type-species: *Ephemerella doddsi* Needham, monotypy. Ephemerellidae: Ephemerellinae: Ephemerellini: Ephemerellae. Subgenus of *Drunella* Needham.
*Eatonella* was originally proposed as a subgenus of *Ephemerella* Walsh. The species included in *Eatonella* were not explictly stated in the original description. *Ephemerella doddsi* was the only included species as determined by process of elimination.

## Genus *Eatonia* Ali

*Eatonia* Ali, 1970:121. Type-species: *Eatonia khyberensis* Ali, monotypy. Preoccupied. Oligoneuriidae: Isonychiinae. Synonym of *Isonychia* Eaton (Hubbard & Peters, 1978:31).

## Genus *Eatonia* Bengtsson

*Eatonia* Bengtsson, 1904:131. Type-species: *Eatonia borealis* Bengtsson [nomen nudum], monotypy. Preoccupied. Nomen nudum. Siphlonuridae: Siphlonurinae. Synonym of *Parameletus* Bengtsson (Bengtsson, 1930:13).

## Genus **Eatonica** Navás

*Eatonica* Navás, 1913:181. Type-species: *Ephemera schoutedeni* Navás, original designation. Ephemeridae.

## Genus **Eatonigenia** Ulmer

*Eatonigenia* Ulmer, 1939:477. Type-species: *Hexagenia chaperi* Navás, original designation. Ephemeridae.

## Genus *Ecdyonuroides* Dang

*Ecdyonuroides* Dang, 1967:160. Type-species: *Ecdyonurus sumatranus* Ulmer, original designation. Heptageniidae: Heptageniinae. Synonym of *Thalerosphyrus* Eaton.

## Genus **Ecdyonurus** Eaton

*Ecdyonurus* Eaton, 1868b:142. Type-species: *Ephemera venosa* Fabricius, original designation. Heptageniidae: Heptageniinae. Synonyms *Afghanurus* Demoulin, *Akkarion* Flowers, *Cinygmina* Kimmins, *Ecdyurus* Eaton, *Leucrocuta* Flowers, *Nixe* Flowers, *Notacanthurus* Tshernova, and *Paracinygmula* Bajkova.

## Genus *Ecdyurus* Eaton

*Ecdyurus* Eaton, 1868a:90. Type-species: *Ephemera venosa* Fabricius, objective synonymy. Unnecessary replacement name for *Ecdyonurus* Eaton. Heptageniidae: Heptageniinae. Objective synonym of *Ecdyonurus* Eaton.

## Genus **Echinobaetis** Mol

*Echinobaetis* Mol, 1989:61. Type-species: *Echinobaetis phagas* Mol, original designation. Baetidae.

## Genus **Ecuaphlebia** Dominguez

*Ecuaphlebia* Dominguez, 1988:227. Type species: *Ecuaphlebia rumignaui* Dominguez, original designation. Leptophlebiidae: Atalophlebiinae.

## Genus **Edmundsius** Day

*Edmundsius* Day, 1953:19. Type-species: *Edmundsius agilis* Day, original designation. Siphlonuridae: Siphlonurinae.

## Genus **Edmundsula** Sivaramakrishnan

*Edmundsula* Sivaramakrishnan, 1985:235. Type-species: *Edmundsula lotica* Sivaramakrishnan, original designation. Leptophlebiidae: Atalophlebiinae.

## Genus **Elassoneuria** Eaton

*Elassoneuria* Eaton, 1881a:191. Type-species: *Oligoneuria trimeniana* McLachlan, original designation. Oligoneuriidae: Oligoneuriinae. Two subgenera are recognized: *Elassoneuria* (s.s.) and *Madeconeuria* Demoulin.

## Genus **Electrogena** Zurwerra & Tomka

*Electrogena* Zurwerra & Tomka, 1985:102. Type-species: *Baetis lateralis* Curtis, original designation. Heptageniidae: Heptageniinae.

## Genus **Electrogenia** Demoulin

*Electrogenia* Demoulin, 1956a:95. Type-species: *Electrogenia dewalschei* Demoulin, original designation. Heptageniidae: Arthropleinae. *Electrogenia* is known only from fossils.

## Genus *Eopolymitarcys* Tshernova

*Eopolymitarcys* Tshernova, 1934:240. Type-species: *Eopolymitarcys nigridorsum* Tshernova, original designation. Polymitarcyidae: Polymitarcyinae. Synonym of *Ephoron* Williamson (Edmunds & Traver, 1954:239).

## Genus *Epeiron* Demoulin

*Epeiron* Demoulin, 1964b:358. Type-species: *Epeiron amseli* Demoulin, original designation. Heptageniidae: Heptageniinae. Synonym of *Epeorus* Eaton (Tshernova & Belov, 1982:194).

## Genus **Epeorella** Ulmer

*Epeorella* Ulmer, 1939:578. Type-species: *Epeorella borneonia* Ulmer, original designation. Heptageniidae: Heptageniinae.

## Genus **Epeoromimus** Tshernova

*Epeoromimus* Tshernova, 1969:156.   Type-species: *Epeoromimus*
*kazlauskasi* Tshernova, original designation.   Epeoromimidae.
*Epeoromimus* is known only from fossils.

## Genus **Epeorus** Eaton

*Epeorus* Eaton, 1881b:26.   Type-species: *Epeorus torrentium* Eaton,
original designation.   Heptageniidae: Heptageniinae.   Synonym
*Epeiron* Demoulin.

Two subgenera are recognized: *Epeorus* (s.s.) and *Ironopsis* Traver. *Epeorus* has
been treated as a subgenus of both *Heptagenia* Walsh and *Iron* Eaton.

## Genus **Ephemera** Linnaeus

*Ephemera* Linnaeus, 1758:546.   Type-species: *Ephemera vulgata*
Linnaeus, subsequent designation by Westwood, 1836:439.   Ephemeridae.   Synonym *Nirvius* Navás.

Two subgenera are recognized: *Ephemera* (s.s.) and *Aethephemera* McCafferty &
Edmunds.

## Genus **Ephemerella** Walsh

*Ephemerella* Walsh, 1863b:377.   Type-species:*Ephemerella excrucians*
Walsh, subsequent designation by Eaton, 1868a:87.   Ephemerellidae:
Ephemerellinae: Ephemerellini: Ephemerellae.   Synonym *Chitono-
phora* Bengtsson.

Two subgenera are recognized: *Ephemerella* (s.s.) and *Uracanthella* Belov.

## Genus **Ephemerellina** Lestage

*Ephemerellina* Lestage, 1924b:346.   Type-species: *Ephemerellina*
*barnardi* Lestage, original designation.   Ephemerellidae: Teloganodinae.

Three subgenera are recognized: *Ephemerellina* (s.s.), *Austremerella* Riek, and
*Lithogloea* Barnard.

## Genus **Ephemeropsis** Eichwald

*Ephemeropsis* Eichwald, 1864:21.   Type-species: *Ephemeropsis*
*trisetalis* Eichwald, monotypy.   Hexagenitidae.   Synonym *Phacelobranchus* Handlirsch.

*Ephemeropsis* is known only from fossils.

## Genus **Ephemerythus** Gillies
*Ephemerythus* Gillies, 1960:35.   Type-species: *Ephemerythus niger* Gillies, original designation.   Tricorythidae: Ephemerythinae. Two subgenera are recognized: *Ephemerythus* (s.s.) and *Tricomerella* Demoulin.

## Genus **Ephoron** Williamson
*Ephoron* Williamson, 1802:71.   Type-species: *Ephoron leukon* Williamson, monotypy.   Polymitarcyidae:  Polymitarcyinae.   Synonyms *Eopolymitarcys* Tshernova and *Polymitarcys* Eaton.

## Genus *Esbenophlebia* Lestage
*Esbenophlebia* Lestage, 1924b:336.   Type-species: *Adenophlebia westermanni* Esben-Petersen, original designation.  Leptophlebiidae: Atalophlebiinae.   Synonym of *Adenophlebia* Eaton (Barnard, 1932: 240).

## Genus *Esca* Sellards
*Esca* Sellards, 1909:354.   Type-species: *Therates planus* Sellards, objective  synonymy.    Replacement  name  for  *Therates*  Sellards (preoccupied).
    Although it was originally proposed as an Ephemeroptera under the name *Therates* Sellards, placement of *Esca* is uncertain, but it is certainly not an Ephemeroptera.

## Genus *Eucharidis* Joly & Joly
*Eucharidis* Joly & Joly, 1877:314.  Type-species: *Eucharidis reamurii* Joly & Joly, monotypy.   Potamanthidae.   Synonym of *Potamanthus* Pictet (Eaton, 1883:79).

## Genus *Eudoter* Tillyard
*Eudoter* Tillyard, 1936:443. Type-species: *Eudoter delicatulus* Tillyard, original designation. Misthodotidae. Synonym of *Misthodotes* Sellards (Carpenter, 1980:237).

## Genus *Euphlebia* Crass
*Euphlebia* Crass, 1947b:104.  Type-species: *Euphlebia bicolor* Crass, monotypy. Leptophlebiidae: Atalophlebiinae. Synonym of *Adenophlebiodes* Ulmer (Edmunds, 1953:79).

## Genus *Euphyurus* Bengtsson
*Euphyurus* Bengtsson, 1909:4. Type-species: *Euphyurus albitarsis* Bengtsson, monotypy. Leptophlebiidae: Leptophlebiinae. Synonym of *Leptophlebia* Westwood (Lestage, 1917:333).

## Genus *Eurycaenis* Bengtsson
*Eurycaenis* Bengtsson, 1917:186. Type-species: *Brachycercus harrisellus* Curtis, original designation. Caenidae. Objective synonym of *Brachycercus* Curtis.

## Genus **Eurylophella** Tiensuu
*Eurylophella* Tiensuu, 1935:20. Type-species: *Eurylophella karelica* Tiensuu, monotypy. Ephemerellidae: Ephemerellinae: Ephemerellini: Timpanogae. Synonym *Melanameletus* Tiensuu.

## Subgenus **Euthraulus** Barnard
*Euthraulus* Barnard, 1932:249. Type-species: *Euthraulus elegans* Barnard, monotypy. Leptophlebiidae: Atalophlebiinae. Synonym *Thraululus* Ulmer. Subgenus of *Choroterpes* Eaton.

## Genus **Euthyplocia** Eaton
*Euthyplocia* Eaton, 1871:67. Type-species: *Palingenia hecuba* Hagen, monotypy. Euthyplociidae: Euthyplociinae.

## Genus **Exeuthyplocia** Lestage
*Exeuthyplocia* Lestage, 1918:74. Type-species: *Euthyplocia minima* Ulmer, monotypy. Euthyplociidae: Exeuthyplociinae.

## Genus **Fallceon** Waltz & McCafferty
*Fallceon* Waltz & McCafferty, 1987c:668. Type-species: *Baetis quilleri* Dodds, original designation. Baetidae.

## Genus **Farrodes** Peters
*Farrodes* Peters, 1971:5. Type-species: *Farrodes hyalinus* Peters, original designation. Leptophlebiidae: Atalophlebiinae.

## Subgenus **Fascioculus** Pescador & Berner
*Baetisca* (*Fascioculus*) Pescador & Berner, 1981:168 (Type-species: *Baetisca escambiensis* Berner, original designation). Baetiscidae.

## Genus **Fittkaulus** Savage & Peters

*Fittkaulus* Savage & Peters, 1978:293. Type-species: *Fittkaulus maculatus* Savage & Peters, original designation. Leptophlebiidae: Atalophlebiinae.

## Genus **Foliomimus** Sinitshenkova

*Foliomimus* Sinitshenkova, 1985:21. Type species: *Epeoromimus tertius* Tshernova, original designation. Epeoromimidae. *Foliomimus* is known only from fossils.

## Genus *Fontainica* McCafferty

*Fontainica* McCafferty, 1968:293. Type-species: *Fontainica josettae* McCafferty, monotypy. Palingeniidae: Palingeniinae. Synonym of *Cheirogenesia* Demoulin (McCafferty & Edmunds, 1976:189).

## Genus **Fulleta** Navás

*Fulleta* Navás, 1930:318. Type-species: *Fulleta dentata* Navás, original designation. Leptophlebiidae: Atalophlebiinae.

## Genus **Fulletomimus** Demoulin

*Fulletomimus* Demoulin, 1956b:280. Type-species: *Fulletomimus marlieri* Demoulin, original designation. Leptophlebiidae: Atalophlebiinae.

## Genus **Garinjuga** Campbell & Suter

*Garinjuga* Campbell & Suter, 1988:262. Type-species: *Garinjuga maryannae* Campbell & Suter, original designation. Leptophlebiidae: Atalophlebiinae

## Genus **Gilliesia** Peters & Edmunds

*Gilliesia* Peters & Edmunds, 1970:189. Type-species: *Thraulus hindustanicus* Gillies, original designation. Leptophlebiidae: Leptophlebiinae.

## Genus **Guajirolus** Flowers

*Guajirolus* Flowers, 1985:27. Type-species: *Guajirolus ektrapeloglossa* Flowers, original designation. Baetidae.

## Subgenus **Guayakia** Dominguez & Flowers

*Hermanella* (*Guayakia*) Dominguez & Flowers, 1989:563.   Type-species: *Thraulus maculipennis* Ulmer, original designation.   Leptophlebiidae: Atalophlebiinae.

## Genus **Habroleptoides** Schoenemund

*Habroleptoides* Schoenemund, 1929:222.  Type-species: *Potamanthus modestus* Hagen, monotypy.  Leptophlebiidae: Leptophlebiinae.

## Genus **Habrophlebia** Eaton

*Habrophlebia* Eaton, 1881a:195.  Type-species: *Ephemera fusca* Curtis, original designation.  Leptophlebiidae: Leptophlebiinae.

Two subgenera are recognized: *Habrophlebia* (s.s.) and *Hesperaphlebia* Peters.

## Genus **Habrophlebiodes** Ulmer

*Habrophlebiodes* Ulmer, 1920b:39.  Type-species: *Habrophlebia americana* Banks, original designation.  Leptophlebiidae: Leptophlebiinae.

## Genus **Hagenulodes** Ulmer

*Hagenulodes* Ulmer, 1920b:37.  Type-species: *Hagenulodes braueri* Ulmer, original designation.  Leptophlebiidae: Atalophlebiinae.

## Genus **Hagenulopsis** Ulmer

*Hagenulopsis* Ulmer, 1920b:34.  Type-species: *Hagenulopsis diptera* Ulmer, original designation.  Leptophlebiidae: Atalophlebiinae.

## Genus **Hagenulus** Eaton

*Hagenulus* Eaton, 1882:207.  Type-species: *Hagenulus caligatus* Eaton, original designation.  Leptophlebiidae: Atalophlebiinae.

## Genus *Haplobaetis* Navás

*Haplobaetis* Navás, 1922b:115.  Type-species: *Haplobaetis umbratus* Navás, original designation.  Baetidae.  Synonym of *Centroptiloides* Lestage (Lestage, 1945:89).

## Genus *Haplogenia* Blair

*Haplogenia* Blair, 1929:254.  Type-species: *Haplogenia southi* Blair, monotypy.  Heptageniidae: Arthropleinae.  Synonym of *Arthroplea* Bengtsson (Bengtsson, 1930:27).

## Genus **Haplohyphes** Allen
*Haplohyphes* Allen, 1966:566. Type-species: *Haplohyphes huallaga* Allen, original designation. Leptohyphidae: Leptohyphinae.

## Genus **Hapsiphlebia** Peters & Edmunds
*Hapsiphlebia* Peters & Edmunds, 1972:1401. Type-species: *Atalophlebia anastomosis* Demoulin, original designation. Leptophlebiidae: Atalophlebiinae.

## Genus **Harpagobaetis** Mol
*Harpagobaetis* Mol, 1986:63. Type-species: *Harpagobaetis gulosus* Mol, original designation. Baetidae.

## Genus **Heptagenia** Walsh
*Heptagenia* Walsh, 1863a:197. Type-species: *Palingenia flavescens* Walsh, subsequent designation by Eaton, 1868a:90. Heptageniidae: Heptageniinae. Synonym *Sigmoneuria* Demoulin.

Three subgenera of *Heptagenia* are recognized: *Heptagenia* (s.s.) Walsh, *Dacnogenia* Kluge, and *Kageronia* Matsumura. The subgenus *Heptagenia* (*Parastenacron*) was proposed by Kluge in Braasch & Soldán (1988) but is not available because it was published without a description.

## Genus **Hermanella** Needham & Murphy
*Hermanella* Needham & Murphy, 1924:39. Type-species: *Hermanella thelma* Needham & Murphy, original designation. Leptophlebiidae: Atalophlebiinae.

## Genus **Hermanellopsis** Demoulin
*Hermanella* (*Hermanellopsis*) Demoulin, 1955a:8. Type-species: *Hermanella incertans* Spieth, original designation. Leptophlebiidae: Atalophlebiinae.

*Hermanellopsis* was originally described as a subgenus of *Hermanella* Needham & Murphy.

## Subgenus **Hesperaphlebia** Peters
*Habrophlebia* (*Hesperaphlebia*) Peters, 1979:53. Type-species: *Habrophlebia vibrans* Needham, original designation. Leptophlebiidae: Leptophlebiinae. Subgenus of *Habrophlebia* Eaton.

## Genus **Heterocloeon** McDunnough
*Heterocloeon* McDunnough, 1925:175. Type-species: *Centroptilum curiosum* McDunnough, original designation. Baetidae. Synonym *Rheobaetis* Müller-Liebenau.

*Heterocloeon* has been treated as a synonym of *Baetis* Leach.

## Genus **Heterogenesia** Dang

*Heterogenesia* Dang, 1967:159.   Type-species: *Heterogenesia chinei* Dang, original designation.   Palingeniidae: Palingeniinae.

## Genus **Hexagenia** Walsh

*Hexagenia* Walsh, 1863a:197.   Type-species: *Ephemera limbata* Serville, subsequent designation by Eaton, 1868a:85.   Ephemeridae.
Two subgenera are recognized: *Hexagenia* (s.s.) and *Pseudeatonica* Spieth.

## Genus **Hexagenites** Scudder

*Hexagenites* Scudder, 1880:6.   Type-species: *Hexagenites weyenberghii* Scudder, original designation.   Hexagenitidae.   Synonyms *Paedephemera* Handlirsch and *Stenodicranum* Demoulin.
*Hexagenites* is known only from fossils.

## Genus **Hexameropsis** Tshernova & Sinitshenkova

*Hexameropsis* Tshernova & Sinitshenkova, 1974:132.   Type-species: *Hexameropsis selini* Tshernova & Sinitshenkova, original designation. Hexagenitidae.
*Hexameropsis* is known only from fossils.

## Subgenus *Holobaetis* Sukatskene

*Baetis* (*Holobaetis*) Sukatskene, 1962:111.   Not available, Art. 13b, no type designated.   Baetidae.   Subgenus of *Baetis* Leach.

## Genus **Homoeoneuria** Eaton

*Homoeoneuria* Eaton, 1881a:192.   Type-species: *Homoeoneuria salvinae* Eaton, original designation.   Oligoneuriidae: Oligoneuriinae.
Two subgenera are recognized: *Homoeoneuria* (s.s.) and *Notachora* Pescador & Peters.

## Subgenus **Homoleptohyphes** Allen & Murvosh

*Tricorythodes* (*Homoleptohyphes*) Allen & Murvosh, 1987:39.   Type-species: *Tricorythodes dimorphus* Allen, original designation. Leptohyphidae: Leptohyphinae.   Subgenus of *Tricorythodes* Ulmer.

## Genus **Homothraulus** Demoulin

*Homothraulus* Demoulin, 1955a:11.   Type-species: *Thraulus misionensis* Esben-Petersen, original designation.   Leptophlebiidae: Atalophlebiinae.

## Genus **Huizhougenia** Lin
*Huizhougenia* Lin, 1980:216. Type species: *Huizhougenia orbicularis* Lin, original designation. Family incertus. *Huizhougenia* is known only from fossils.

## Subgenus **Hyalophlebia** Demoulin
*Adenophlebiodes (Hyalophlebia)* Demoulin, 1955h:287. Type-species: *Atalophlebia dentifera* Navás, original designation. Leptophlebiidae: Atalophlebiinae. Subgenus of *Adenophlebiodes* Ulmer.

## Genus **Hylister** Dominguez & Flowers
*Hylister* Dominguez & Flowers, 1989:565. Type-species: *Hylister plaumanni* Dominguez & Flowers, original designation. Leptophlebiidae: Atalophlebiinae.

## Genus **Hyrtanella** Allen & Edmunds
*Hyrtanella* Allen & Edmunds, 1976:133. Type-species: *Hyrtanella christineae* Allen & Edmunds, original designation. Ephemerellidae: Ephemerellinae: Hyrtanellini.

## Genus **Ichthybotus** Eaton
*Ichthybotus* Eaton, 1899:285. Type-species: *Ephemera hudsoni* McLachlan, monotypy. Ephemeridae.

## Genus **Indialis** Peters & Edmunds
*Indialis* Peters & Edmunds, 1970:208. Type-species: *Indialis badia* Peters & Edmunds, original designation. Leptophlebiidae: Atalophlebiinae.

## Genus **Indobaetis** Müller-Liebenau & Morihara
*Indobaetis* Müller-Liebenau & Morihara, 1982:26. Type-species: *Indobaetis costai* Müller-Liebenau & Morihara, original designation. Baetidae.

## Genus **Indocloeon** Müller-Liebenau
*Indocloeon* Müller-Liebenau, 1982:125. Type-species: *Indocloeon primum* Müller-Liebenau, original designation. Baetidae.

## Genus **Insulibrachys** Soldán

*Insulibrachys* Soldán, 1986:333. Type-species: *Insulibrachys needhami* Soldán, original designation. Caenidae.

## Genus **Iron** Eaton

*Iron* Eaton, 1883:pl. 23-24. Type-species: *Epeorus longimanus* Eaton, monotypy. Heptageniidae: Heptageniinae.

*Iron* has been treated as both a synonym and a subgenus of *Epeorus* Eaton.

## Genus **Ironodes** Traver

*Ironodes* Traver, 1935:32. Type-species: *Iron nitidus* Eaton, original designation. Heptageniidae: Heptageniinae.

*Ironodes* has been treated as both a synonym and a subgenus of *Epeorus* Eaton.

## Subgenus **Ironopsis** Traver

*Ironopsis* Traver, 1935:36. Type-species: *Iron grandis* McDunnough, original designation. Heptageniidae: Heptageniinae. Subgenus of *Epeorus* Eaton.

## Genus **Isca** Gillies

*Isca* Gillies, 1951:127. Type-species: *Isca purpurea* Gillies, original designation. Leptophlebiidae: Atalophlebiinae.

Three subgenera are recognized: *Isca* (s.s.), *Minyphlebia* Peters & Edmunds, and *Tanycola* Peters & Edmunds.

## Genus **Isonychia** Eaton

*Isonychia* Eaton, 1871:134. Type-species: *Isonychia manca* Eaton, original designation. Oligoneuriidae: Isonychiinae. Synonyms *Chirotonetes* Eaton, *Eatonia* Ali [not available], and *Jolia* Eaton.

Three subgenera are recognized: *Isonychia* (s.s.), *Borisonychia* McCafferty, and *Prionodes* Kondratieff & Voshell.

## Genus **Isothraulus** Towns & Peters

*Isothraulus* Towns & Peters, 1979:439. Type-species: *Isothraulus abditus* Towns & Peters, original designation. Leptophlebiidae: Atalophlebiinae.

## Genus **Jappa** Harker

*Jappa* Harker, 1954:257. Type-species: *Jappa kutera* Harker, original designation. Leptophlebiidae: Atalophlebiinae.

## Genus **Jarmila** Demoulin
*Jarmila* Demoulin, 1970b:7. Type-species: *Jarmila elongata* Demoulin, original designation. Jarmilidae.
*Jarmila* is known only from fossils.

## Genus *Jolia* Eaton
*Jolia* Eaton, 1881a:192. Type-species: *Palingenia roeselli* Joly, original designation. Oligoneuriidae: Isonychiniinae. Synonym of *Isonychia* Eaton (Needham, 1905:28).
*Jolia* has been treated as a synonym of *Chirotonetes* Eaton.

## Genus **Jubabaetis** Müller-Liebenau
*Jubabaetis* Müller-Liebenau, 1980:103. Type-species: *Jubabaetis pescadori* Müller-Liebenau, original designation. Baetidae.

## Subgenus **Kageronia** Matsumura
*Kageronia* Matsumura, 1931:1479. Type-species: *Kageronia suzukiella* Matsumura, monotypy. Heptageniidae: Heptageniinae. Subgenus of *Heptagenia* Walsh.

## Genus *Kamia* Martynov
*Kamia* Martynov, 1928:4. Type-species: Kamia *augustovenosa* Martynov, original designation.
Originally proposed as a mayfly, the placement of *Kamia* is uncertain, although it is probably a Paleodictyoptera.

## Genus **Kariona** Peters & Peters
*Kariona* Peters & Peters, 1981b:245. Type-species: *Kariona quintata* Peters & Peters, original designation. Leptophlebiidae: Atalophlebiinae.

## Genus **Kimminsula** Peters & Edmunds
*Kimminsula* Peters & Edmunds, 1970:192. Type-species: *Potamanthus annulatus* Hagen, original designation. Leptophlebiidae: Atalophlebiinae.

## Genus **Kirrara** Harker
*Kirrara* Harker, 1954:259. Type-species: *Kirrara procera* Harker, original designation. Leptophlebiidae: Atalophlebiinae.

## Genus **Koorrnonga** Campbell & Suter

*Koorrnonga* Campbell & Suter, 1988:267.  Type-species: *Leptophlebia inconspicua* Eaton, original designation.  Leptophlebiidae: Atalophlebiinae

## Genus **Kukalova** Demoulin

*Kukalova* Demoulin, 1970b:6.  Type-species: *Kukalova americana* Demoulin, original designation.  Protereismatidae.  *Kukalova* is known only from fossils.

## Subgenus **Labiobaetis** Novikova & Kluge

*Baetis* (*Labiobaetis*) Novikova & Kluge, 1987:13.  Type-species: *Baetis atrebatinus* Eaton, original designation.  Baetidae.  Subgenus of *Baetis* Leach.

## Genus **Lachlania** Hagen

*Lachlania* Hagen, 1868:372.  Type-species: *Lachlania abnormis* Hagen, original designation.  Oligoneuriidae: Oligoneuriinae.  Synonyms *Alloydia* Needham, *Neophlebia* Navás, *Noya* Navás, and *Noyopsis* Navás.

## Subgenus **Languidipes** Hubbard

*Povilla* (*Languidipes*) Hubbard, 1984:29.  Type-species: *Asthenopus corporaali* Lestage, original designation.  Polymitarcyidae: Asthenopodinae.  Subgenus of *Povilla* Navás.

## Genus **Leentvaaria** Demoulin

*Leentvaaria* Demoulin, 1966a:13.  Type-species: *Leentvaaria palpalis* Demoulin, original designation.  Leptophlebiidae: Atalophlebiinae.

## Genus **Lepegenia** Peters, Peters, & Edmunds

*Lepegenia* Peters, Peters, & Edmunds, 1978:109.  Type-species: *Lepegenia lineata* Peters, Peters, & Edmunds, original designation.  Leptophlebiidae: Atalophlebiinae.

## Genus **Lepeorus** Peters, Peters, & Edmunds

*Lepeorus* Peters, Peters, & Edmunds, 1978:100.  Type-species: *Lepeorus goyi* Peters, Peters, & Edmunds, original designation.  Leptophlebiidae: Atalophlebiinae.

## Genus **Lepismophlebia** Demoulin
*Lepismophlebia* Demoulin, 1968a:7. Type-species: *Lepisma platymera* Scudder, original designation. Leptophlebiidae: Subfamily incertus. *Lepismophlebia* is known only from fossils.

## Genus **Leptohyphes** Eaton
*Leptohyphes* Eaton, 1882:208. Type-species: *Leptohyphes eximius* Eaton, monotypy. Leptohyphidae: Leptohyphinae. Synonym *Bruchella* Navás.

## Genus **Leptohyphodes** Ulmer
*Leptohyphodes* Ulmer, 1920b:50. Type-species: *Potamanthus inanis* Pictet, original designation. Leptohyphidae: Leptohyphinae.

## Genus **Leptoneta** Sinitshenkova
*Leptoneta* Sinitshenkova, 1989:35. Type-species: *Leptoneta calyptrata* Sinitshenkova, original designation. Leptophlebiidae: Mesonetinae. *Leptoneta* is known only from fossils.

## Genus **Leptophlebia** Westwood
*Leptophlebia* Westwood, 1840:31. Type-species: *Ephemera vespertina* Linnaeus, monotypy. Leptophlebiidae: Leptophlebiinae. Synonyms *Blasturus* Westwood and *Euphyurus* Bengtsson.

## Genus **Lestagella** Demoulin
*Lestagella* Demoulin, 1970d:130. Type-species: *Lithogloea penicillata* Barnard, original designation. Ephemerellidae: Teloganodinae.

## Genus **Leucorhoenanthus** Lestage
*Leucorhoenanthus* Lestage, 1931a:134. Type-species: *Rhoenanthus macedonicus* Ulmer, original designation. Neoephemeridae. Synonym *Caenomera* Demoulin.
*Leucorhoenanthus* has been treated as a synonym of *Oreianthus* Traver and a subgenus of *Neoephemera* McDunnough.

## Genus *Leucrocuta* Flowers
*Leucrocuta* Flowers, 1980:297. Type-species: *Heptagenia maculipennis* McDunnough, original designation. Heptageniidae: Heptageniinae. Synonym of *Ecdyonurus* Eaton (Tshernova et al., 1986:114).

## Genus **Liebebiella** Waltz & McCafferty

*Liebebiella* Waltz & McCafferty, 1987d:556. Type-species: *Pseudo-cloeon orientale* Müller-Liebenau, original designation. Baetidae.

## Subgenus **Lithogloea** Barnard

*Lithogloea* Barnard, 1932:252. Type-species: *Lithogloea harrisoni* Barnard, monotypy. Ephemerellidae: Teloganodinae. Subgenus of *Ephemerellina* Lestage.

## Genus **Lithoneura** Carpenter

*Lithoneura* Carpenter, 1938:446. Type species: *Lithoneura lameerei* Carpenter, original designation. Syntonopteridae.

*Lithoneura* is known only from fossils.

## Genus **Litobrancha** McCafferty

*Litobrancha* McCafferty, 1971:45. Type-species: *Hexagenia recurvata* Morgan, monotypy. Ephemeridae.

## Genus *Litophlebia* Hubbard & Riek

*Litophlebia* Hubbard & Riek, 1977:260. Type-species: *Xenophlebia optata* Riek, objective synonymy. Replacement name for *Xenophlebia* Riek.

Originally proposed as a mayfly (under the name *Xenophlebia* Riek), *Litophlebia* belongs to the Megasecoptera.

## Genus *Longinella* Gros & Lestage

*Longinella* Gros & Lestage, 1927:158. Type-species: *Euthyplocia guntheri* Navás, original designation. Euthyplociidae: Euthyplociinae. Synonym of *Campylocia* Needham & Murphy (Ulmer, 1932:205).

## Genus *Loxophlebia* Martynov

*Loxophlebia* Martynov, 1928:8. Type-species: *Loxophlebia apicalis* Martynov, original designation. Protereismatidae. Synonym of *Protereisma* Sellards (Rohdendorf, 1957:76).

## Subgenus **Maccaffertium** Bednarik

*Stenonema (Maccaffertium)* Bednarik, 1979:191. Type-species: *Heptagenia integer* McDunnough, original designation. Heptageniidae: Heptageniinae. Subgenus of *Stenonema* Traver.

## Genus **Macdunnoa** Lehmkuhl
*Macdunnoa* Lehmkuhl, 1979:860. Type-species: *Macdunnoa nipawinia* Lehmkuhl, original designation. Heptageniidae: Heptageniinae.

## Genus **Machadorythus** Demoulin
*Machadorythus* Demoulin, 1959:251. Type-species: *Machadorythus palanquim* Demoulin, monotypy. Tricorythidae: Machadorythinae.

### Subgenus **Madeconeuria** Demoulin
*Elasoneuria* (*Madeconeuria*) Demoulin, 1973:8. Type-species: *Elassoneuria insulicola* Demoulin, original designation. Oligoneuriidae: Oligoneuriinae. Subgenus of *Elassoneuria* Eaton.

## Genus **Magallanella** Pescador & Peters
*Magallanella* Pescador & Peters, 1980b:336. Type-species: *Magallanella flinti* Pescador & Peters, original designation. Leptophlebiidae: Atalophlebiinae.

## Genus **Maheathraulus** Peters, Gillies, & Edmunds
*Maheathraulus* Peters, Gillies, & Edmunds, 1964:122. Type-species: *Hagenulus scotti* Eaton, original designation. Leptophlebiidae: Atalophlebiinae.

## Genus **Manohyphella** Allen
*Manohyphella* Allen, 1973:160. Type-species: *Manohyphella keiseri* Allen, original designation. Ephemerellidae: Teloganodinae.

## Genus *Masharikella* Peters, Gillies, & Edmunds
*Masharikella* Peters, Gillies, & Edmunds, 1964:118. Type-species: *Hagenulus fasciatus* Kimmins, original designation. Leptophlebiidae: Atalophlebiinae. Synonym of *Thraulus* Eaton (Peters & Edmunds, 1970:203).

## Genus **Massartella** Lestage
*Massartella* Lestage, 1930b:251. Type-species: *Atalophlebia brieni* Lestage, original designation. Leptophlebiidae: Atalophlebiinae.
*Massartella* has been considered a synonym of *Atalophlebia* Eaton.

## Genus **Massartellopsis** Demoulin
*Massartellopsis* Demoulin, 1955c:9.　Type-species: *Massartellopsis irarrazavali* Demoulin, original designation.　Leptophlebiidae: Atalophlebiinae.

## Genus **Matsumuracloeon** Hubbard
*Matsumuracloeon* Hubbard, 1989:388. Type-species: *Pseudocloeon aino* Matsumura, monotypy.　Replaecement name for *Pseudocloeon* Matsumura [not available].　Baetidae.

## Genus **Mauiulus** Towns & Peters
*Mauiulus* Towns & Peters, 1979a:224.　Type-species: *Mauiulus luma* Towns & Peters, original designation.　Leptophlebiidae: Atalophlebiinae.

## Subgenus **Mayobaetis** Waltz & McCafferty
*Moribaetis* (*Mayobaetis*) Waltz & McCafferty, 1985:240.　Type-species: *Baetis ellenae* Mayo, original designation.　Baetidae.　Subgenus of *Moribaetis* Waltz & McCafferty.

## Genus *Mecus* Sellards
*Mecus* Sellards, 1909:151.　Type-species: *Scopus gracilis* Sellards, objective synonymy.　Replacement name for *Scopus* Sellards. Protereismatidae.　Synonym of *Protereisma* Sellards (Tillyard, 1932:244).

## Genus **Megaglena** Peters & Edmunds
*Megaglena* Peters & Edmunds, 1970:210.　Type-species: *Megaglena brincki* Peters & Edmunds, original designation.　Leptophlebiidae: Atalophlebiinae.

## Genus *Melanameletus* Tiensuu
*Melanameletus* Tiensuu, 1935:15.　Type-species: *Melanameletus brunnescens* Tiensuu, monotypy.　Ephemerellidae: Ephemerellinae: Ephemerellini: Timpanogae.　Synonym of *Eurylophella* Tiensuu (Edmunds & Traver, 1954:238).

## Genus **Melanemerella** Ulmer
*Melanemerella* Ulmer, 1920b:43. Type-species: *Melanemerella brasiliana* Ulmer, original designation. Ephemerellidae: Melanemerellinae.

## Genus **Meridialaris** Peters and Edmunds
*Meridialaris* Peters and Edmunds, 1972:1405. Type-species: *Deleatidium laminatum* Ulmer, original designation. Leptophlebiidae: Atalophlebiinae.

## Genus **Mesephemera** Handlirsch
*Mesephemera* Handlirsch, 1906:600. Type-species: *Ephemera procera* Hagen, subsequent designation by Hubbard, 1981:69. Mesephemeridae.
*Mesephemera* is known only from fossils.

## Genus **Mesobaetis** Brauer, Redtenbacher, & Ganglbauer
*Mesobaetis* Brauer, Redtenbacher, & Ganglbauer, 1889:5. Type-species: *Mesobaetis sibiricus* Brauer, Redtenbacher, & Ganglbauer, monotypy. Baetidae.
*Mesobaetis* is known only from fossils.

## Genus **Mesogenesia** Tshernova
*Mesogenesia* Tshernova, 1977:92. Type-species: *Mesogenesia petersae* Tshernova, original designation. Palingeniidae: Palingeniinae.
*Mesogenesia* is known only from fossils.

## Genus **Mesoneta** Brauer, Redtenbacher, & Ganglbauer
*Mesoneta* Brauer, Redtenbacher, & Ganglbauer, 1889:4. Type-species: *Mesoneta antiqua* Brauer, Redtenbacher, & Ganglbauer, monotypy. Leptophlebiidae: Mesonetinae.
*Mesoneta* is known only from fossils.

## Genus *Mesonetopsis* Ping
*Mesonetopsis* Ping, 1935:112. Type-species: *Mesonetopsis zingi* Ping, original designation.
Originally proposed as a mayfly, *Mesonetopsis* probably belongs to the Odonata.

## Genus **Mesopalingea** Whalley & Jarzembowski
*Mesopalingea* Whalley & Jarzembowski, 1985:384.   Type-species:
*Mesopalingea lerida* Whalley & Jarzembowski.   Palingeniidae.
*Mesopalingea* is known only from fossils.

## Genus **Mesoplectopteron** Handlirsch
*Mesoplectopteron* Handlirsch, 1918:112.   Type-species: *Mesoplectop-
teron longipes* Handlirsch, original designation.   Mesoplectopteridae.
*Mesoplectopteron* is known only from fossils.

## Genus **Mesoplocia** Demoulin
*Mesoplocia* Demoulin, 1952c:11.   Type-species: *Mesoplocia intermedia*
Demoulin, original designation.   Euthyplociidae: Euthyplociinae.

## Genus **Metamonius** Eaton
*Metamonius* Eaton, 1885:208.   Type-species: *Siphlurus anceps* Eaton,
original designation.   Siphlonuridae: Siphlonurinae.

## Genus *Metreletus* Demoulin
*Metreletus* Demoulin, 1951:10.   Type-species: *Metretopus goetghebueri*
Lestage,   original   designation.      Siphlonuridae:   Siphlonurinae.
Synonym of *Ameletus* Eaton (Jacob, 1984:182).

## Genus **Metretopus** Eaton
*Metretopus* Eaton, 1901:253.   Type-species: *Metretopus norvegicus*
Eaton, original designation.   Metretopodidae.

## Genus *Metreturus* Burks
*Metreturus* Burks, 1953:146.   Type-species: *Metreturus pecatonica*
Burks, original designation.   Siphlonuridae: Acanthametropodinae.
Synonym of *Acanthametropus* Tshernova (Edmunds & Allen, 1957:
318).

## Genus **Microphlebia** Savage & Peters
*Microphlebia* Savage & Peters, 1983:569.   Type-species: *Microphlebia
surinamensis* Savage & Peters, original designation.   Leptophlebiidae:
Atalophlebiinae.

## Subgenus **Minyphlebia** Peters & Edmunds
*Isca (Minyphlebia)* Peters & Edmunds, 1970:217. Type-species: *Isca janiceae* Peters & Edmunds, original designation. Leptophlebiidae: Atalophlebiinae. Subgenus of *Isca* Gillies.

## Genus **Miocoenogenia** Tshernova
*Miocoenogenia* Tshernova, 1962:944. Type-species: *Miocoenogenia gorbunovi* Tshernova, original designation. Heptageniidae: Heptageniinae.
*Miocoenogenia* is known only from fossils.

## Genus **Mirawara** Harker
*Mirawara* Harker, 1954:261. Type-species: *Mirawara aapta* Harker, original designation. Ameletopsidae: Ameletopsinae.

## Genus **Miroculis** Edmunds
*Miroculis* Edmunds, 1963:34. Type-species: *Miroculis rossi* Edmunds, original designation. Leptophlebiidae: Atalophlebiinae.
Four subgenera are recognized: *Miroculis* (s.s.), *Atroari* Savage & Peters, *Ommaethus* Savage & Peters, and *Yaruma* Savage & Peters.

## Genus **Miroculitus** Savage & Peters
*Miroculitus* Savage & Peters, 1983:566. Type-species: *Choroterpes emersoni* Needham & Murphy, original designation. Leptophlebiidae: Atalophlebiinae.

## Genus **Misthodotes** Sellards
*Misthodotes* Sellards, 1909:151. Type-species: *Dromeus obtusus* Sellards, objective synonymy. Misthodotidae. Replacement name for *Dromeus* Sellards. Synonyms *Dromeus* Sellards and *Eudoter* Tillyard.
*Misthodotes* is known only from fossils.

## Genus **Mogzonurella** Sinitshenkova
*Mogzonurella* Sinitshenkova, 1985:16. Type species: *Mogzonurella dissimilis* Sinitshenkova, original designation. Oligoneuriidae: Coloburiscinae.
*Mogzonurella* is known only from fossils.

## Genus **Mogzonurus** Sinitshenkova
*Mogzonurus* Sinitshenkova, 1985:15.    Type species: *Mogzonurus elevatus* Sinitshenkova, original designation.  Oligoneuriidae: Coloburiscinae.
*Mogzonurus* is known only from fossils.

## Genus **Mongologenites** Sinitshenkova
*Mongologenites* Sinitshenkova, 1986:45.  Type-species: *Mongologenites laqueatus* Sinitshenkova, original designation.  Hexagenitidae.
*Mongologenites* is known only from fossils.

## Subgenus *Montobaetis* Kazlauskas
*Baetis* (*Montobaetis*) Kazlauskas, 1972:338.    Not available, Art. 13b, no type designated.  Baetidae.  Subgenus of *Baetis* Leach.

## Genus **Moribaetis** Waltz & McCafferty
*Moribaetis* Waltz & McCafferty, 1985:240.    Type-species: *Baetis maculipennis* Flowers, original designation.  Baetidae.
Two subgenera are recognized: *Moribaetis* (s.s.) and *Mayobaetis* Waltz & McCafferty.

## Genus **Mortogenesia** Lestage
*Mortogenesia* Lestage, 1923:110. Type-species: *Palingenia mesopotamica* Morton, original designation.  Palingeniidae: Palingeniinae.

## Genus **Murphyella** Lestage
*Murphyella* Lestage, 1930a:439. Type-species: *Murphyella needhami* Lestage, original designation.    Oligoneuriidae: Coloburiscinae.
Synonym *Dictyosiphlon* Lestage.

## Genus **Mutelocloeon** Gillies & Elouard
*Mutelocloeon* Gillies & Elouard, 1990:289. Type-species: *Mutelocloeon bihoumi* Gillies & Elouard, original designation.  Baetidae.

## Subgenus **Myllonella** Allen
*Drunella* (*Myllonella*) Allen, 1980:80.    Type-species: *Ephemerella coloradensis* Dodds, original designation. Ephemerellidae: Ephemerellinae: Ephemerellini: Ephemerellae.  Subgenus of *Drunella* Needham.

## Genus **Nathanella** Demoulin
*Nathanella* Demoulin, 1955d:1.    Type-species: *Nathanella indica* Demoulin, original designation.  Leptophlebiidae: Atalophlebiinae.

## Genus **Neboissophlebia** Dean
*Neboissophlebia* Dean, 1988:39. Type-species: *Neboissophlebia hamulata* Dean, original designation. Leptophlebiidae: Atalophlebiinae.

## Genus **Needhamella** Dominguez & Flowers
*Needhamella* Dominguez & Flowers, 1989:568. Type-species: *Thraulus ehrhardti* Ulmer, original designation. Leptophlebiidae: Atalophlebiinae.

## Genus *Needhamocoenis* Lestage
*Needhamocoenis* Lestage, 1945:85. Type-species: *Caenopsis fugitans* Needham, objective synonymy. Replacement name for *Caenopsis* Needham. Leptohyphidae: Leptohyphinae. Objective synonym of *Tricorythafer* Lestage.

## Subgenus *Neobaetiella* Müller-Liebenau
*Neobaetiella* Müller-Liebenau, 1985:103. Type-species: *Neobaetiella uenoi* Müller-Liebenau, original designation. Baetidae. Synonym of *Baetis (Baetiella)* Uéno (Waltz & McCafferty, 1987d:561).

## Genus *Neobaetis* Navás
*Neobaetis* Navás, 1924:69. Type-species: *Neobaetis paulinus* Navás, original designation. Baetidae. Synonym of *Callibaetis* Eaton (Edmunds, 1974:289).

## Subgenus **Neochoroterpes** Allen
*Choroterpes (Neochoroterpes)* Allen, 1974:163. Type-species: *Choroterpes mexicanus* Allen, original designation. Leptophlebiidae: Atalophlebiinae. Subgenus of *Choroterpes* Eaton.

## Genus *Neocloeon* Traver
*Neocloeon* Traver, 1932:365. Type-species: *Neocloeon alamance* Traver, original designation. Baetidae. Synonym of *Cloeon* Leach (Edmunds, Jensen, & Berner, 1976:173).

## Genus **Neoephemera** McDunnough
*Neoephemera* McDunnough, 1925:168. Type-species: *Neoephemera bicolor* McDunnough, original designation. Neoephemeridae.
Two subgenera are recognized: *Neoephemera* (s.s.) and *Orieanthus* Traver.

## Genus **Neoephemeropsis** Ulmer

*Neoephemeropsis* Ulmer, 1939:483.   Type-species: *Neoephemeropsis caenoides* Ulmer, original designation.   Neoephemeridae.

## Genus **Neohagenulus** Traver

*Neohagenulus* Traver, 1938:8.   Type-species: *Neohagenulus julio* Traver, original designation.   Leptophlebiidae: Atalophlebiinae.

## Genus *Neophlebia* Navás

*Neophlebia* Navás, 1912a:746.   Type-species: *Neophlebia garciai* Navás, monotypy.   Preoccupied.   Oligoneuriidae: Oligoneuriinae. Objective synonym of *Noya* Navás [= *Lachlania* Hagen (Ulmer, 1943:35)].

Navás (1912a) established the new genus *Neophlebia*.   In the same paper, recognizing that the generic name *Neophlebia* was preoccupied, he changed the name to *Noya*.

## Genus **Neopotamanthodes** Hsu

*Neopotamanthodes* Hsu, 1938:221.   Type-species: *Neopotamanthodes lanchi* Hsu, original designation.   Potamanthidae.

## Genus **Neopotamanthus** Wu & You

*Neopotamanthus* Wu & You, 1986:401.   Type-species: *Neopotamanthus youi* Wu & You, original designation.   Potamnthidae.

## Genus **Neozephlebia** Penniket

*Zephlebia* (*Neozephlebia*) Penniket, 1961:9.   Type-species: *Baetis scita* Walker, original designation.   Leptophlebiidae: Atalophlebiinae.

*Neozephlebia* was originally described as a subgenus of *Zephlebia* Penniket.

## Genus **Nesameletus** Tillyard

*Nesameletus* Tillyard, 1933:11.   Type-species: *Chirotonetes ornatus* Eaton, original designation.   Siphlonuridae: Siphlonurinae.

## Genus **Nesophlebia** Peters & Edmunds

*Nesophlebia* Peters & Edmunds, 1964:248.   Type-species: *Nesophlebia adusta* Peters & Edmunds, original designation.   Leptophlebiidae: Atalophlebiinae.

## Genus **Nesoptiloides** Demoulin
*Nesoptiloides* Demoulin, 1973:2. Type-species: *Nesoptiloides intermedia* Demoulin, original designation. Baetidae.

## Genus **Neurocaenis** Navás
*Neurocaenis* Navás, 1936:365. Type-species: *Neurocaenis fuscata* Navás, original designation. Tricorythidae: Tricorythinae.

## Subgenus **Nigrobaetis** Kazlauskas
*Baetis* (*Nigrobaetis*) Kazlauskas in Novikova & Kluge, 1987:8. Type-species: *Ephemera nigra* Linnaeus, original designation. Baetidae. Subgenus of *Baetis* Leach.
*Nigrobaetis* was originally proposed as a genus without a designated type-species (nomen nudum) by Kazlauskas in 1972 (p:338).

## Genus *Nirvius* Navás
*Nirvius* Navás, 1922a:56. Type-species: *Nirvius punctatus* Navás, original designation. Ephemeridae. Synonym of *Ephemera* Linnaeus (Lestage, 1922:253).

## Genus *Nixe* Flowers
*Nixe* Flowers, 1980:299. Type-species: *Ecdyonurus lucidipennis* Clemens, original designation. Heptageniidae: Heptageniinae. Synonym of *Ecdyonurus* Eaton (Tshernova et al., 1986:114).

## Genus *Notacanthurus* Tshernova
*Notacanthurus* Tshernova, 1974:812. Type-species: *Ecdyonurus zhiltzovae* Tshernova, original designation. Heptageniidae: Heptageniinae. Synonym of *Ecdyonurus* Eaton (Tshernova et al., 1986:114).

## Genus **Notachalcus** Peters & Peters
*Notachalcus* Peters & Peters, 1981a:233. Type-species: *Notachalcus corbassoni* Peters & Peters, original designation. Leptophlebiidae: Atalophlebiinae.

## Subgenus **Notachora** Pescador & Peters
*Homoeoneuria* (*Notachora*) Pescador & Peters, 1980a:385. Type-species: *Homoeoneuria fittkaui* Pescador & Peters, original designation. Oligoneuriidae: Oligoneuriinae. Subgenus of *Homoeoneuria* Eaton.

## Subgenus **Notobaetis** Morihara & Edmunds

*Notobaetis* Morihara & Edmunds, 1980:606. Type-species: *Notobaetis penai* Morihara & Edmunds, original designation. Baetidae. Subgenus of *Cloeodes* Traver.
*Notobaetis* has been treated as a synonym of *Cloeodes* Traver.

## Genus *Notonurus* Crass

*Notonurus* Crass, 1947a:126. Type-species: *Notonurus cooperi* Crass, original designation. Heptageniidae: Heptageniinae. Synonym of *Compsoneuria* Eaton (Gillies, 1984:21; Braasch & Soldán, 1986a:46).
*Notonurus* was synonymyzed by Gillies (1984) with *Compsoneuriella* Ulmer which is a synonym of *Compsoneuria* Eaton (Braasch & Soldán, 1986a).

## Genus **Notophlebia** Peters & Edmunds

*Notophlebia* Peters & Edmunds, 1970:213. Type-species: *Notophlebia hyalina* Peters & Edmunds, original designation. Leptophlebiidae: Atalophlebiinae.

## Genus **Nousia** Navás

*Nousia* Navás, 1918:213. Type-species: *Nousia delicata* Navás, original designation. Leptophlebiidae: Atalophlebiinae. Synonym *Atalonella* Needham & Murphy.

## Genus *Noya* Navás

*Noya* Navás, 1912a:746. Type-species: *Neophlebia garciai* Navás, objective synonymy. Replacement name for *Neophlebia* Navás. Oligoneuriidae: Oligoneuriinae. Synonym of *Lachlania* Hagen (Ulmer, 1943:35).

## Genus *Noyopsis* Navás

*Noyopsis* Navás, 1924:70. Type-species: *Noyopsis fusca* Navás, original designation. Oligoneuriidae: Oligoneuriinae. Synonym of *Lachlania* Hagen (Ulmer, 1943:35).

## Genus **Nyungara** Dean

*Nyungara* Dean, 1987:91. Type-species: *Nyungara bunni* Dean, original designation. Leptophlebiidae: Atalophlebiinae.

## Genus **Oboriphlebia** Hubbard & Kukalová-Peck

*Oboriphlebia* Hubbard & Kukalová-Peck, 1980:29. Type-species: *Kukalova moravica* Demoulin, original designation. Oboriphlebiidae. *Oboriphlebia* is known only from fossils.

## Genus **Olgisca** Demoulin

*Olgisca* Demoulin, 1970c:6. Type-species: *Paedephemera schwertschlageri* Handlirsch, original designation. Siphlonuridae: Siphlonurinae. *Olgisca* is known only from fossils.

## Genus **Oligoneuria** Pictet

*Oligoneuria* Pictet, 1843:288. Type-species: *Oligoneuria anomala* Pictet, monotypy. Oligoneuriidae: Oligoneuriinae.

## Genus **Oligoneuriella** Ulmer

*Oligoneuriella* Ulmer, 1924a:31. Type-species: *Oligoneuria rhenana* Imhoff, original designation. Oligoneuriidae: Oligoneuriinae.

## Genus **Oligoneurioides** Demoulin

*Oligoneurioides* Demoulin, 1955a:24. Type-species: *Oligoneurioides amazonicus* Demoulin, original designation. Oligoneuriidae: Oligoneuriinae.

## Genus **Oligoneuriopsis** Crass

*Oligoneuriopsis* Crass, 1947b:52. Type-species: *Oligoneuriopsis lawrencei* Crass, monotypy. Oligoneuriidae: Oligoneuriinae.

## Genus **Oligoneurisca** Lestage

*Oligoneurisca* Lestage, 1938b:318. Type-species: *Oligoneuriella borysthenica* Tshernova, monotypy. Oligoneuriidae: Oligoneuriinae.

## Genus *Oligophlebia* Demoulin

*Oligophlebia* Demoulin, 1965a:146. Type-species: *Oligophlebia calliarcys* Demoulin, original designation. Leptophlebiidae: Leptophlebiinae. Synonym of *Paraleptophlebia* Lestage (Demoulin, 1970a:11).

## Subgenus **Ommaethus** Savage & Peters

*Miroculis* (*Ommaethus*) Savage & Peters, 1983:559. Type-species: *Miroculis mourei* Savage & Peters, original designation. Leptophlebiidae: Atalophlebiinae. Subgenus of *Miroculis* Edmunds.

## Genus **Oniscigaster** McLachlan
*Oniscigaster* McLachlan, 1873:109. Type-species: *Oniscigaster wakefieldi* McLachlan, monotypy. Oniscigastridae.

## Genus **Ophelmatostoma** Waltz & McCafferty
*Ophelmatostoma* Waltz & McCafferty, 1987a:96. Type-species: *Ophelmatostoma kimminsi* Waltz & McCafferty, original designation. Baetidae.

## Genus *Ordella* Campion
*Ordella* Campion, 1923:518. Type-species: *Caenis macrura* Stephens, original designation. Caenidae. Objective synonym of *Caenis* Stephens.

## Subgenus **Oreianthus** Traver
*Oreianthus* Traver, 1931:104. Type-species: *Oreianthus purpureus* Traver, original designation. Neoephemeridae. Subgenus of *Neoephemera* McDunnough.
    *Orieanthus* has been treated as a synonym of *Neoephemera* McDunnough.

## Genus **Ororotsia** Traver
*Ororotsia* Traver, 1939:33. Type-species: *Ororotsia hutchinsoni* Traver, original designation. Heptageniidae: Heptageniinae.

## Genus **Ounia** Peters & Peters
*Ounia* Peters & Peters, 1981a:240. Type-species: *Ounia loisoni* Peters & Peters, original designation. Leptophlebiidae: Atalophlebiinae.

## Genus *Oxycypha* Burmeister
*Oxycypha* Burmeister, 1839:796. Type-species: *Oxycypha luctuosa* Burmeister, subsequent designation by Jacob, 1974:96. Caenidae. Synonym of *Caenis* Stephens (Jacob, 1974:96).
    *Oxycypha* has been treated as a synonym of *Brachycercus* Curtis.

## Genus *Paedephemera* Handlirsch
*Paedephemera* Handlirsch, 1906:601. Type-species: *Ephemera multinervosa* Oppenheim, subsequent designation by Demoulin, 1954e:571. Hexagenitidae. Synonym of *Hexagenites* Scudder (Demoulin, 1970c:5).

## Genus **Paegniodes** Eaton
*Paegniodes* Eaton, 1881b:23. Type-species: *Heptagenia cupulata* Eaton, original designation. Heptageniidae: Heptageniinae.

## Genus **Palaeobaetodes** Brito
*Palaeobaetodes* Brito, 1987:594. Type-species: *Palaeobaetodes costalimai* Brito, original designation. Baetidae.

## Genus *Paleoameletus* Lestage
*Paleoameletus* Lestage, 1940:124. Type-species: *Ameletus primitivus* Traver. original designation. Siphlonuridae: Siphlonurinae. Synonym of *Ameletus* Eaton (Edmunds & Traver, 1954:237).

## Genus **Palinephemera** Lin
*Palinephemera* Lin, 1986:26. Type-species: *Palinephemera densivena* Lin, original designation. Mesephemeridae.
*Palinephemera* is known only from fossils.

## Genus **Palingenia** Burmeister
*Palingenia* Burmeister, 1839:802. Type-species: *Ephemera longicauda* Olivier [as Swammerdam], subsequent designation by Hagen in Walsh, 1863a:173. Palingeniidae: Palingeniinae.

## Genus **Palingeniopsis** Martynov
*Palingeniopsis* Martynov, 1932:10. Type-species: *Palingeniopsis praecox* Martynov, original designation. Mesephemeridae.
*Palingeniopsis* is known only from fossils.

## Genus *Palmenia* Aro
*Palmenia* Aro, 1910:28. Type-species: *Palmenia fennica* Aro, monotypy. Siphlonuridae: Siphlonurinae. Synonym of *Parameletus* Bengtsson (Bengtsson, 1930:13; see Hubbard, 1977:409).

## Genus **Papposa** Peters & Peters
*Papposa* Peters & Peters, 1981b:250. Type-species: *Papposa hirsuta* Peters & Peters, original designation. Leptophlebiidae: Atalophlebiinae.

## Genus **Parabaetis** Haupt

*Parabaetis* Haupt, 1956:32.    Type-species: *Parabaetis eocaenicus*
Haupt, original designation.    Ephemeridae.

*Parabaetis* is known only from fossils.

## Genus *Paracinygmula* Bajkova

*Paracinygmula* Bajkova, 1975:54.    Type-species: *Paracinygmula*
*zhilzovae* Bajkova, original designation.    Heptageniidae: Heptageniinae.    Synonym of *Ecdyonurus* Eaton (Tshernova, 1978:541).

## Genus **Paracloeodes** Day

*Paracloeodes* Day, 1955:121.    Type-species: *Paracloeodes abditus* Day,
original designation.    Baetidae.

## Genus **Paraleptophlebia** Lestage

*Paraleptophlebia* Lestage, 1917:340.    Type-species: *Ephemera cincta*
Retzius, subsequent designation by Lestage, 1924b:42.    Leptophlebiidae: Leptophlebiinae.    Synonym *Oligophlebia* Demoulin.

## Genus **Parameletus** Bengtsson

*Parameletus* Bengtsson, 1908:242.    Type-species: *Parameletus chelifer*
Bengtsson, monotypy.    Siphlonuridae: Siphlonurinae.    Synonyms
*Eatonia* Bengtsson [not available], *Palmenia* Aro, *Potameis* Bengtsson,
*Siphlonuroides* McDunnough, and *Sparrea* Esben-Petersen.

*Parameletus* was originally proposed as a *nomen nudum* by Bengtsson (1904:131).

## Subgenus *Parastenacron* Kluge

*Heptagenia* (*Parastenacron*) Kluge in Braasch & Soldán, 1988:119.
Type-species: *Ephemera fuscogrisea* Retzius, monotypy.    Heptageniidae: Heptageniinae.    Not available.    Subgenus of *Heptagenia* Walsh.

*Heptagenia* (*Parastenacron*) is not available because it was published without an
associated description.

## Genus **Peloracantha** Peters & Peters

*Peloracantha* Peters & Peters, 1980:70.    Type-species: *Peloracantha*
*titan* Peters & Peters, original designation.    Leptophlebiidae: Atalophlebiinae.

# Genus **Penaphlebia** Peters & Edmunds
*Penaphlebia* Peters & Edmunds, 1972:1399. Type-species: *Atalophlebia chilensis* Eaton, original designation. Leptophlebiidae: Atalophlebiinae.

# Genus **Penniketellus** Towns & Peters
*Penniketellus* Towns & Peters, 1979b:449. Type-species: *Penniketellus insolitus* Towns & Peters, original designation. Leptophlebiidae: Atalophlebiinae.

# Genus **Pentagenia** Walsh
*Pentagenia* Walsh, 1863a:196. Type-species: *Palingenia vittigera* Walsh, subsequent designation by Eaton, 1868a:85. Palingeniidae: Pentageniinae.

# Genus *Perissophlebia* Savage
*Perissophlebia* Savage, 1982:209. Type-species: *Perissophlebia flinti* Savage, original designation. Preoccupied. Leptophlebiidae: Atalophlebiinae. Objective synonym of *Perissophlebiodes* Savage.

# Genus **Perissophlebiodes** Savage
*Perissophlebiodes* Savage, 1983:204. Type-species: *Perissophlebia flinti* Savage, objective synonymy. Leptophlebiidae: Atalophlebiinae. Replacement name for *Perissophlebia* Savage.

# Genus **Petersophlebia** Demoulin
*Petersophlebia* Demoulin, 1973:12. Type-species: *Petersophlebia insularis* Demoulin, original designation. Leptophlebiidae: Atalophlebiinae.

# Genus **Petersula** Sivaramakrishnan
*Petersula* Sivaramakrishnan, 1984:194. Type-species: *Petersula courtallensis* Sivaramakrishnan, original designation. Leptophlebiidae: Atalophlebiinae.

# Genus *Phacelobranchus* Handlirsch
*Phacelobranchus* Handlirsch, 1906:604. Type-species: *Phacelobranchus braueri* Handlirsch, monotypy. Hexagenitidae. Synonym of *Ephemeropsis* Eichwald (Tshernova, 1961:861).

Genus **Philolimnias** Hong
*Philolimnias* Hong, 1979:336.   Type-species: *Philolimnias sinica* Hong,
original designation.   Family incertus.
*Philolimnias* is known only from fossils.

Genus **Phthartus** Handlirsch
*Phthartus* Handlirsch, 1904:6.   Type-species: *Phthartus rossicus*
Handlirsch, original designation.   Protereismatidae.
*Phthartus* is known only from fossils.

Genus *Pinctodia* Sellards
*Pinctodia* Sellards, 1907:352.   Type-species: *Pinctodia curta* Sellards,
original designation.   Protereismatidae.   Synonym of *Protereisma*
Sellards (Tillyard, 1932:244).

Genus **Platybaetis** Müller-Liebenau
*Platybaetis* Müller-Liebenau, 1980:104.   Type-species: *Platybaetis
edmundsi* Müller-Liebenau, original designation.   Baetidae.

Genus **Plethogenesia** Ulmer
*Plethogenesia* Ulmer, 1920a:102.   Type-species: *Palingenia papuana*
Eaton, monotypy.   Palingeniidae: Palingeniinae.   Synonym *Tritogen-
esia* Lestage.

Genus *Polymitarcys* Eaton
*Polymitarcys* Eaton, 1868a:84.   Type-species: *Ephemera virgo* Olivier,
original designation.   Polymitarcyidae: Polymitarcyinae.   Synonym of
*Ephoron* Williamson (Spieth, 1940:109).

Genus **Polyplocia** Lestage
*Polyplocia* Lestage, 1921:212.   Type-species: *Polyplocia vitalisi* Lestage,
original designation.   Euthyplociidae: Euthyplociinae.
*Polyploycia* was described as a new genus a second time by Lestage (1924a:80).

Genus **Polythelais** Demoulin
*Polythelais* Demoulin, 1973:15.   Type-species: *Polythelais digitata*
Demoulin, original designation.   Leptophlebiidae: Atalophlebiinae.

Genus **Potamanthellus** Lestage
*Potamanthellus* Lestage, 1931a:120. Type-species: *Potamanthellus horai* Lestage, original designation. Neoephemeridae. Synonym *Rhoenanthodes* Lestage.

Genus **Potamanthindus** Lestage
*Potamanthindus* Lestage, 1931a:123. Type-species: *Potamanthindus auratus* Lestage, original designation. Potamanthidae.

Genus **Potamanthodes** Ulmer
*Potamanthodes* Ulmer, 1920b:11. Type-species: *Potamanthus formosus* Eaton, original designation. Potamanthidae.

Genus **Potamanthus** Pictet
*Potamanthus* Pictet, 1843:197. Type-species: *Ephemera lutea* Linnaeus, subsequent designation by Eaton, 1868a:86. Potamanthidae. Synonym *Eucharidis* Joly & Joly.

Genus *Potameis* Bengtsson
*Potameis* Bengtsson, 1909:13. Type-species: *Potameis elegans* Bengtsson, subsequent designation by Hubbard, 1979a:412. Siphlonuridae: Siphlonurinae. Synonym of *Parameletus* Bengtsson (Bengtsson, 1930:13).

Genus **Povilla** Navás
*Povilla* Navás, 1912c:402. Type-species: *Povilla adusta* Navás, original designation. Polymitarcyidae: Asthenopodinae.
Two subgenera are recognized: *Povilla* (s.s.) and *Languidipes* Hubbard.

Genus **Poya** Peters & Peters
*Poya* Peters & Peters, 1980:65. Type-species: *Poya brunnea* Peters & Peters, original designation. Leptophlebiidae: Atalophlebiinae.

Subgenus **Prionodes** Kondrateiff & Voshell
*Isonychia* (*Prionodes*) Kondrateiff & Voshell, 1983:129. Type-species: *Isonychia georgiae* McDunnough, original designation. Oligoneuriidae: Isonychiinae. Subgenus of *Isonychia* Eaton.

## Genus **Proameletus** Sinitshenkova

*Proameletus* Sinitshenkova, 1976:86.   Type-species: *Proameletus caudatus* Sinitshenkova, original designation.   Siphlonuridae: Siphlonurinae.
*Proameletus* is known only from fossils.

## Genus **Proboscidoplocia** Demoulin

*Proboscidoplocia* Demoulin, 1966b:946.   Type-species: *Euthyplocia sikorai* Vayssière, original designation.   Euthyplociidae: Euthyplociinae.

## Genus **Procloeon** Bengtsson

*Procloeon* Bengtsson, 1915:34.   Type-species: *Cloeon bifidum* Bengtsson, objective synonymy.   Baetidae.   Replacement name for *Pseudocloeon* Bengtsson.
*Procloeon* has been treated as both a synonym and a subgenus of *Baetis* Leach

## Genus *Procloeon* Matsumura

*Procloeon* Matsumura, 1931:1472.   Not available, Art. 13b, no type designated.   Preoccupied.   Baetidae.   Synonym of *Promatsumura* Hubbard (Hubbard, 1988:240).

## Genus *Prodromites* Cockerell

*Prodromites* Cockerell, 1924:136.   Type-species: *Prodromus rectus* Sellards, objective synonymy.   Replacement name for *Prodromus* Sellards.   Protereismatidae.   Synonym of *Protereisma* Sellards (Tillyard, 1932:244).

## Genus *Prodromus* Sellards

*Prodromus* Sellards, 1907:349.   Type-species: *Prodromus rectus* Sellards, original designation.   Preoccupied.   Protereismatidae.   Objective synonym of *Prodromites* Cockerell [= *Protereisma* Sellards (Tillyard, 1932:244)].

## Genus **Promatsumura** Hubbard

*Promatsumura* Hubbard, 1988:240.   Type-species: *Procloeon nipponicum* Matsumura, original designation.   Replacement name for *Procloeon* Matsumura [not available].

## Genus **Promirara** Jell & Duncan

*Promirara* Jell & Duncan, 1986:116. Type-species: *Promirara cephalota* Jell & Duncan, original designation. Siphlonuridae. *Promirara* is known only from fossils.

## Genus *Propalingenia* Handlirsch

*Propalingenia* Handlirsch, 1906:86. Type-species: *Palingenia feistmantelli* Fritsch, monotypy. Originally proposed as an Ephemeroptera, *Propalingenia* belongs in the Paleodictyoptera.

## Genus **Prosopistoma** Latreille

*Prosopistoma* Latreille, 1833:33. Type-species: *Prosopistoma variegatum* Latreille, subsequent designation by Eaton, 1884:150. Prosopistomatidae. Synonyms *Binoculis* Geoffroy [not available] and *Chelysentomon* Joly & Joly.

## Genus *Protechma* Sellards

*Protechma* Sellards, 1907:349. Type-species: *Protechma acuminatum* Sellards, original designation. Protereismatidae. Synonym of *Protereisma* Sellards (Tillyard, 1932:244).

## Genus **Protereisma** Sellards

*Protereisma* Sellards, 1907:347. Type-species: *Protereisma permianum* Sellards, original designation. Protereismatidae. Synonyms *Bantiska* Sellards, *Loxophlebia* Martynov, *Mecus* Sellards, *Pinctodia* Sellards, *Prodromites* Cockerell, *Prodromus* Sellards, *Protechma* Sellards, *Recter* Sellards, *Rekter* Sellards, and *Scopus* Sellards. *Protereisma* is known only from fossils.

## Genus **Protobehningia** Tshernova

*Protobehningia* Tshernova in Tshernova & Bajkova, 1960:411. Type-species: *Protobehningia asiatica* Tshernova, original designation. Behningiidae.

## Genus **Protoligoneuria** Demoulin

*Protoligoneuria* Demoulin, 1955g:271. Type-species: *Protoligoneuria limai* Demoulin, original designation. Oligoneuriidae: Oligoneuriinae. *Protoligoneuria* is known only from fossils.

## Subgenus **Psammonella** Glazaczow

*Pseudocentroptiloides (Psammonella)* Glazaczow in Jacob & Glazaczow, 1986:203. Type-species: *Pseudocentroptiloides ceylonica* Glazaczow, original designation. Subgenus of *Pseudocentroptiloides* Jacob. Baetidae.

## Subgenus **Pseudeatonica** Spieth

*Hexagenia (Pseudeatonica)* Spieth, 1941:269. Type-species: *Hexagenia mexicana* Eaton, original designation. Ephemeridae. Subgenus of *Hexagenia* Walsh.

Pseudeatonica has been treated as a subgenus of *Eatonica* Navás.

## Genus **Pseudiron** McDunnough

*Pseudiron* McDunnough, 1931:91. Type-species: *Pseudiron centralis* McDunnough, original designation. Heptageniidae: Anepeorinae.

## Genus *Pseudocaenis* Soldán

*Pseudocaenis* Soldán, 1978:124. Type-species: *Pseudocaenis queenslandica* Soldán, original designation. Caenidae. Synonym of *Tasmanocoenis* Lestage (Suter, 1984:105).

## Genus **Pseudocentroptiloides** Jacob

*Pseudocentroptiloides* Jacob in Jacob & Glazaczow, 1986:198. Type-species: *Pseudocentroptilum shadini* Kazlauskas, original designation. Baetidae.

Two subgenera are recognized: *Pseudocentroptiloides* (s.s.) and *Psammonella* Glazaczow.

## Genus **Pseudocentroptilum** Bogoesco

*Pseudocentroptilum* Bogoesco, 1947:602. Type-species: *Pseudocentroptilum motasi* Bogoesco, monotypy. Baetidae. Synonym *Cloeoptilum* Kazlauskas [not available].

## Genus **Pseudocloeon** Klapálek

*Pseudocloeon* Klapálek, 1905:105. Type-species: *Pseudocloeon kraepelini* Klapálek, monotypy. Baetidae.

## Genus *Pseudocloeon* Bengtsson

*Pseudocloeon* Bengtsson, 1914:218. Type-species: *Cloeon bifidum* Bengtsson, original designation. Preoccupied. Baetidae. Objective synonym of *Procloeon* Bengtsson (Bengtsson, 1915:34).

## Genus *Pseudocloeon* Matsumura

*Pseudocloeon* Matsumura, 1931:1473. Type-species: *Pseudocloeon aino* Matsumura, monotypy. Preoccupied. Baetidae. Objective synonym of *Matsumuracloeon* Hubbard (Hubbard, 1989:388).

## Genus *Pseudoligoneuria* Ulmer

*Pseudoligoneuria* Ulmer, 1939:540. Type-species: *Pseudoligoneuria feuerborni* Ulmer, original designation. Oligoneuriidae: Chromarcyinae. Synonym of *Chromarcys* Navás (Demoulin, 1967:4).

## Genus *Pseudopalingenia* Handlirsch

*Pseudopalingenia* Handlirsch, 1906:124. Type-species: *Palingenia feistmantelli* Handlirsch, original designation.
Originally proposed as an Ephemeroptera, *Pseudopalingenia* belongs to the Paleodictyoptera.

## Genus **Pseudopannota** Waltz & McCafferty

*Pseudopannota* Waltz & McCafferty, 1987a:95. Type-species: *Pseudocloeon bertrandi* Demoulin, original designation. Baetidae.

## Subgenus **Pseudulmeritus** Traver

*Ulmeritus* (*Pseudulmeritus*) Traver, 1959:7. Type-species: *Thraulodes flavopedes* Spieth, original designation. Leptophlebiidae: Atalophlebiinae. Subgenus of *Ulmeritus* Traver.

## Genus **Rallidens** Penniket

*Rallidens* Penniket, 1966:164. Type-species: *Rallidens mcfarlanei* Penniket, original designation. Siphlonuridae: Rallidentinae.

## Subgenus **Raptobaetopus** Müller-Liebenau

*Raptobaetopus* Müller-Liebenau, 1978:470. Type-species: *Raptobaetopus orientalis* Müller-Liebenau, original designation. Baetidae. Subgenus of *Baetopus* Keffermüller.

## Genus **Raptoheptagenia** Whiting & Lehmkuhl

*Raptoheptagenia* Whiting & Lehmkuhl, 1987a:405. Type-species: *Heptagenia cruentata* Walsh, original designation. Heptageniidae: Heptageniinae.

## Genus *Recter* Sellards

*Recter* Sellards, 1909:151.    Type-species: *Rekter arcuatus* Sellards, objective synonymy. Replacement name for *Rekter* Sellards. Protereismatidae. Synonym of *Protereisma* Sellards (Tillyard, 1932:244).

## Genus *Rekter* Sellards

*Rekter* Sellards, 1907:349.    Type-species: *Rekter arcuatus* Sellards, original designation. Protereismatidae. Objective synonym of *Recter* Sellards [= *Protereisma* Sellards (Tillyard, 1932:244)].

## Genus *Remipalpus* Bengtsson

*Remipalpus* Bengtsson, 1908:242.    Type-species: *Remipalpus elegans* Bengtsson, monotypy. Heptageniidae: Arthropleinae.    Synonym of *Arthroplea* Bengtsson (Bengtsson, 1930:27).

## Genus *Rheobaetis* Müller-Liebenau

*Rheobaetis* Müller-Liebenau, 1974:555.    Type-species: *Rheobaetis petersi* Müller-Liebenau, original designation. Baetidae. Synonym of *Heterocloeon* McDunnough (McCafferty & Provansha, 1975:124).

## Genus **Rhigotopus** Pescador & Peters

*Rhigotopus* Pescador & Peters, 1982:4.    Type-species: *Rhigotopus andinensis* Pescador & Peters, original designation. Leptophlebiidae: Atalophlebiinae.

## Subgenus **Rhionella** Allen

*Cinctocostella* (*Rhionella*) Allen, 1980:83. Type-species: *Ephemerella insolta* Allen, original designation. Ephemerellidae: Ephemerellinae: Ephemerellini: Ephemerellae. Subgenus of *Cincticostella* Allen.

## Genus **Rhithrocloeon** Gillies

*Rhithrocloeon* Gillies, 1985:13.    Type-species: *Cloeon permirum* Kopelke, original designation. Baetidae.

## Genus **Rhithrogena** Eaton

*Rhithrogena* Eaton, 1881b:23.    Type-species: *Baetis semicoloratus* Curtis, original designation. Heptageniidae: Heptageniinae.

## Genus **Rhithrogeniella** Ulmer
*Rhithrogeniella* Ulmer, 1939:575. Type-species: *Rhithrogeniella ornata* Ulmer, original designation. Heptageniidae: Heptageniinae.

## Subgenus *Rhodobaetis* Kazlauskas
*Baetis* (*Rhodobaetis*) Kazlauskas, 1972:338. Not available, Art. 13b, no type designated. Baetidae. Subgenus of *Baetis* Leach.

## Genus *Rhoenanthodes* Lestage
*Rhoenanthodes* Lestage, 1931a:136. Type-species: *Rhoenanthus amabilis* Eaton, original designation. Neoephemeridae. Synonym of *Potamanthellus* Lestage (Hsu, 1937:137).

## Genus **Rhoenanthopsis** Ulmer
*Rhoenanthopsis* Ulmer, 1932:212. Type-species: *Rhoenanthus magnificus* Ulmer, original designation. Potamanthidae.

## Genus **Rhoenanthus** Eaton
*Rhoenanthus* Eaton, 1881a:192. Type-species: *Rhoenanthus speciosus* Eaton, original designation. Potamanthidae.

## Genus *Scopus* Sellards
*Scopus* Sellards, 1907:352. Type-species: *Scopus gracilis* Sellards, original designation. Protereismatidae. Objective synonym of *Mecus* Sellards [= *Protereisma* Sellards (Tillyard, 1932:244)].

## Genus **Secochela** Pescador & Peters
*Secochela* Pescador & Peters, 1982:7. Type-species: *Secochela illiesi* Pescador & Peters, original designation. Leptophlebiidae: Atalophlebiinae.

## Genus **Serratella** Edmunds
*Ephemerella* (*Serratella*) Edmunds, 1959:544. Type-species: *Ephemerella serrata* Morgan, original designation. Ephemerellidae: Ephemerellinae: Ephemerellini: Ephemerellae.
    *Serratella* was originally proposed as a subgenus of *Ephemerella* Walsh.

## Genus **Siberiogenites** Sinitshenkova
*Siberiogenites* Sinitshenkova, 1985:20. Type species: *Siberiogenites angustatus* Sinitshenkova, original designation. Hexagenitidae.

*Siberiogenites* is known only from fossils. It was originally proposed as a *nomen nudum* by Sinitshenkova (1984:63).

## Genus *Sigmoneuria* Demoulin

*Sigmoneuria* Demoulin, 1964b:353. Type-species: *Sigmoneuria amseli* Demoulin, original designation. Heptageniidae: Heptageniinae. Synonym of *Heptagenia* Walsh (Kluge, 1987:314.)

## Genus **Simothraulopsis** Demoulin

*Simothraulopsis* Demoulin, 1966a:15. Type-species: *Simothraulopsis surinamensis* Demoulin, original designation. Leptophlebiidae: Atalophlebiinae.

## Genus **Simothraulus** Ulmer

*Simothraulus* Ulmer, 1939:508. Type-species: *Simothraulus seminiger* Ulmer, original designation. Leptophlebiidae: Atalophlebiinae.

## Genus *Sinoephemera* Ping

*Sinoephemera* Ping, 1935:111. Type-species: *Sinoephemera kingi* Ping, original designation.

Originally proposed as an Ephemeroptera, *Sinoephemera* probably belongs in the Plecoptera.

## Genus **Siphlaenigma** Penniket

*Siphlaenigma* Penniket, 1962:389. Type-species: *Siphlaenigma janae* Penniket, original designation. Siphlaenigmatidae.

## Genus **Siphlonella** Needham & Murphy

*Siphlonella* Needham & Murphy, 1924:30. Type-species: *Siphlonella ventilans* Needham & Murphy, monotypy. Oniscigastridae.

## Genus **Siphlonisca** Needham

*Siphlonisca* Needham, 1909:72. Type-species: *Siphlonisca aerodromia* Needham, original designation. Siphlonuridae: Siphlonurinae.

## Genus *Siphlonuroides* McDunnough

*Siphlonuroides* McDunnough, 1923:48. Type-species: *Siphlonuroides croesus* McDunnough, original designation. Siphlonuridae: Siphlonurinae. Synonym of *Parameletus* Bengtsson (McDunnough, 1932:81).

## Genus **Siphlonurus** Eaton
*Siphlonurus* Eaton, 1868a:89. Type-species: *Baetis flavidus* Pictet, original designation. Siphlonuridae: Siphlonurinae. Synonyms *Andromina* Navás and *Siphlurus* Eaton.
Two subgenera are recognized: *Siphlonurus* (s.s.) and *Siphlurella* Bengtsson.

## Genus **Siphloplecton** Clemens
*Siphloplecton* Clemens, 1915:258. Type-species: *Siphlurus flexus* Clemens, original designation. Metretopodidae.

## Subgenus **Siphlurella** Bengtsson
*Siphlurella* Bengtsson, 1909:11. Type-species: *Siphlurella thomsoni* Bengtsson, subsequent designation by Hubbard, 1979a:412. Siphlonuridae: Siphlonurinae. Subgenus of *Siphlonurus* Eaton.
*Siphlurella* has been treated as a synonym of *Siphlonurus* Eaton.

## Genus **Siphluriscus** Ulmer
*Siphluriscus* Ulmer, 1920b:61. Type-species: *Siphluriscus chinensis* Ulmer, original designation. Siphlonuridae: Acanthametropodinae.

## Genus **Siphlurites** Cockerell
*Siphlurites* Cockerell, 1923:170. Type-species: *Siphlurites explanatus* Cockerell, original designation. Oligoneuriidae: Coloburiscinae.
*Siphlurites* is known only from fossils.

## Genus *Siphlurus* Eaton
*Siphlurus* Eaton, 1871:125. Type-species: *Baetis flavidus* Pictet, objective synonymy. Unnecessary replacement name for *Siphlonurus* Eaton. Siphlonuridae: Siphlonurinae. Objective synonym of *Siphlonurus* Eaton.

## Genus **Spaniophlebia** Eaton
*Spaniophlebia* Eaton, 1881a:191. Type-species: *Spaniophlebia trailae* Eaton, original designation. Oligoneuriidae: Oligoneuriinae.

## Genus *Sparrea* Esben-Petersen
*Sparrea* Esben-Petersen, 1909:554. Type-species: *Sparrea norvegica* Esben-Petersen, monotypy. Siphlonuridae: Siphlonurinae. Synonym of *Parameletus* Bengtsson (Bengtsson, 1930:13).

## Genus *Spinadis* Edmunds & Jensen

*Spinadis* Edmunds & Jensen, 1974:495.　Type-species: *Spinadis wallacei* Edmunds & Jensen, original designation.　Heptageniidae: Anepeorinae.　Synonym of *Anepeorus* McDunnough (McCafferty & Provansha, 1988:15).

## Genus **Stackelbergisca** Tshernova

*Stackelbergisca* Tshernova, 1967:323.　Type-species: *Stackelbergisca sibirica* Tshernova, original designation.　Siphlonuridae: Acanthametropodinae.

*Stackelbergisca* is known only from fossils.

## Genus **Stenacron** Jensen

*Stenacron* Jensen, 1974:225.　Type-species: *Baetis interpunctatus* Say, original designation.　Heptageniidae: Heptageniinae.

## Genus *Stenodicranum* Demoulin

*Stenodicranum* Demoulin, 1954e:571.　Type-species: *Ephemera cellulosa* Hagen, original designation.　Hexagenitidae.　Synonym of *Hexagenites* Scudder (Demoulin, 1970c:7).

## Genus **Stenonema** Traver

*Stenonema* Traver, 1933:113.　Type-species: *Heptagenia tripunctata* Banks, original designation.　Heptageniidae: Heptageniinae.

Two subgenera are recognized: *Stenonema* (s.s.) and *Maccaffertium* Bednarik.

## Genus **Succinogenia** Demoulin

*Succinogenia* Demoulin, 1965a:151.　Type-species: *Succinogenia larssoni* Demoulin, original designation.　Heptageniidae: Heptageniinae.

*Succinogenia* is known only from fossils.

## Genus **Sulawesia** Peters & Edmunds

*Sulawesia* Peters & Edmunds, 1990:327.　Type-species: *Sulawesia haema* Peters & Edmunds, original designation.　Leptophlebiidae: Atalophlebiinae.

## Genus **Symbiocloeon** Müller-Liebenau

*Symbiocloeon* Müller-Liebenau in Müller-Liebenau & Heard, 1979:57.　Type-species: *Symbiocloeon heardi* Müller-Liebenau, original designation.　Baetidae.

## Genus **Syntonoptera** Handlirsch
*Syntonoptera* Handlirsch, 1911:299.    Type species: *Syntonoptera schucherti* Handlirsch, monotypy.  Syntonopteridae.
*Syntonoptera* is known only from fossils.

## Subgenus **Takobia** Novikova & Kluge
*Baetis* (*Takobia*) Novikova & Kluge, 1987:10.  Type-species: *Centroptilum maxillare* Braasch & Soldán, original designation.  Baetidae. Subgenus of *Baetis* Leach.

## Subgenus **Tanycola** Peters & Edmunds
*Isca* (*Tanycola*) Peters & Edmunds, 1970:219.   Type-species: *Isca serendiba* Peters & Edmunds, original designation.  Leptophlebiidae: Atalophlebiinae.  Subgenus of *Isca* Gillies.

## Genus **Tasmanocoenis** Lestage
*Tasmanocoenis* Lestage, 1931c:53.    Type-species: *Tasmanocoenis tonnoiri* Lestage, original designation.  Caenidae.  Synonym *Pseudocaenis* Soldán.

## Genus **Tasmanophlebia** Tillyard
*Tasmanophlebia* Tillyard, 1921:410.   Type-species: *Tasmanophlebia lacustris* Tillyard, original designation.  Oniscigastridae.  Synonym *Tasmanophlebiodes* Lestage.

## Genus *Tasmanophlebiodes* Lestage
*Tasmanophlebiodes* Lestage, 1935:351.  not available, Art. 13b, no type designated.  Oniscigastridae.  Synonym of *Tasmanophlebia* Tillyard (Riek, 1955:268).

## Genus **Teloganella** Ulmer
*Teloganella* Ulmer, 1939:516.    Type-species: *Teloganella umbrata* Ulmer, original designation.  Ephemerellidae: Teloganodinae.

## Genus **Teloganodes** Eaton
*Teloganodes* Eaton, 1882:208.    Type-species: *Cloe tristis* Hagen, original designation.  Ephemerellidae: Teloganodinae.

Genus **Teloganopsis** Ulmer
*Teloganopsis* Ulmer, 1939:513.    Type-species: *Teloganopsis media*
Ulmer, original designation.    Ephemerellidae: Ephemerellinae:
Ephemerellini: Ephemerellae.
*Teloganopsis* has been treated as a subgenus of *Ephemerella* Walsh.

Subgenus **Terama** Towns
*Zephlebia (Terama)* Towns, 1983:18.    Type-species: *Atalophlebia*
*borealis* Phillips, original designation.    Leptophlebiidae: Atalophlebiinae. Subgenus of *Zephlebia* Penniket.

Genus **Terpides** Demoulin
*Terpides* Demoulin, 1966a:15.    Type-species: *Terpides guyanensis*
Demoulin, original designation.    Leptophlebiidae: Atalophlebiinae.

Genus **Thalerosphyrus** Eaton
*Thalerosphyrus* Eaton, 1881b:22.    Type-species: *Baetis determinatus*
Walker, original designation.    Heptageniidae: Heptageniinae.
Synonym *Ecdyonuroides* Dang.

Genus *Therates* Sellards
*Therates* Sellards, 1907:354.    Type-species: *Therates planus*, original
designation. Preoccupied. Synononym of *Esca* Sellards.
   Originally proposed as a mayfly, *Therates* (the name has been replaced by *Esca*
Sellards) is uncertain, but it is certainly not an Ephemeroptera.

Genus *Thnetus* Handlirsch
*Thnetus* Handlirsch, 1904:7.    Type-species: *Thnetus stuckenbergi*
Handlirsch, original designation.
   Originally proposed as an Ephemeroptera, *Thnetus* probably belongs in the
Paleodictyoptera.

Genus **Thraulodes** Ulmer
*Thraulodes* Ulmer, 1920b:33. Type-species: *Calliarcys laetus* Eaton,
original designation. Leptophlebiidae: Atalophlebiinae.

Genus **Thraulophlebia** Demoulin
*Thraulophlebia* Demoulin, 1955f:227.    Type-species: *Atalophlebia*
*lucida* Ulmer, original designation. Leptophlebiidae: Atalophlebiinae.

## Subgenus *Thraululus* Ulmer

*Thraululus* Ulmer, 1939:499. Type-species: *Thraulus marginatus* Ulmer, original designation. Leptophlebiidae: Atalophlebiinae. Synonym of *Choroterpes* (*Euthraulus*) Barnard (Gillies, 1957:43).

## Genus **Thraulus** Eaton

*Thraulus* Eaton, 1881a:195. Type-species: *Thraulus bellus* Eaton, original designation. Leptophlebiidae: Atalophlebiinae. Synonym *Masharikella* Peters, Gillies, & Edmunds.

## Genus **Timpanoga** Needham

*Ephemerella* (*Timpanoga*) Needham, 1927:115. Type-species: *Ephemerella hecuba* Eaton, monotypy. Ephemerellidae: Ephemerellinae: Ephemerellini: Timpanogae.

*Timpanoga* was originally proposed as a subgenus of *Ephemerella* Walsh.

## Genus **Tindea** Peters & Peters

*Tindea* Peters & Peters, 1980:68. Type-species: *Tindea cochereaui* Peters & Peters, original designation. Leptophlebiidae: Atalophlebiinae.

## Genus **Torephemera** Sinitshenkova

*Torephemera* Sinitshenkova, 1989:40. Type-species: *Torephemera longipes* Sinitshenkova, original designation. Torephemeridae.

*Torephemera* is known only from fossils.

## Genus **Torleya** Lestage

*Torleya* Lestage, 1917:366. Type-species: *Torleya belgica* Lestage, monotypy. Ephemerellidae: Ephemerellinae: Ephemerellini: Ephemerellae.

*Torleya* has been treated as a subgenus of *Ephemerella* Walsh.

## Genus **Tortopus** Needham & Murphy

*Tortopus* Needham & Murphy, 1924:23. Type-species: *Tortopus igaranus* Needham & Murphy, original designation. Polymitarcyidae: Campsurinae.

## Genus **Traverella** Edmunds

*Traverella* Edmunds, 1948:141. Type-species: *Thraulus albertanus* McDunnough, original designation. Leptophlebiidae: Atalophlebiinae.

## Genus **Traverina** Peters

*Traverina* Peters, 1971:9. Type-species: *Traverina cubensis* Peters & Alayo, original designation. Leptophlebiidae: Atalophlebiinae.

## Subgenus **Tribrochella** Allen

*Drunella* (*Tribrochella*) Allen, 1980:80. Type-species: *Ephemerella trispina* Uéno, original designation. Ephemerellidae: Ephemerellinae: Ephemerellini: Ephemerellae. Subgenus of *Drunella* Needham.

## Genus **Trichogenia** Braasch & Soldán

*Trichogenia* Braasch & Soldán, 1988:119. Type-species: *Trichogenia maxillaris* Braasch & Soldán, original designation. Heptageniidae: Heptageniinae.

## Subgenus **Tricomerella** Demoulin

*Ephemerythus* (*Tricomerella*) Demoulin, 1964a:17. Type-species: *Ephemerythus straeleni* Demoulin, monotypy. Tricorythidae: Ephemerythinae. Subgenus of *Ephemerythus* Gillies.

## Subgenus **Tricoryhyphes** Allen & Murvosh

*Tricorythodes* (*Tricoryhyphes*) Allen & Murvosh, 1987:38. Type-species: *Tricorythodes condylus* Allen, original designation. Leptohyphidae: Leptohyphinae. Subgenus of *Tricorythodes* Ulmer.

## Genus **Tricorythafer** Lestage

*Tricorythafer* Lestage, 1942:4. Type-species: *Caenopsis fugitans* Needham, objective synonymy. Leptohyphidae: Leptohyphinae. Replacement name for *Caenopsis* Needham. Synonyms *Caenopsis* Needham and *Needhamocoenis* Lestage.

## Genus **Tricorythodes** Ulmer

*Tricorythodes* Ulmer, 1920b:51. Type-species: *Tricorythus explicatus* Eaton, original designation. Leptohyphidae: Leptohyphinae.

Three subgenera are recognized: *Tricorythodes* (s.s.), *Homoleptohyphes* Allen & Murvosh, and *Tricoryhyphes* Allen & Murvosh.

## Genus **Tricorythopsis** Traver

*Tricorythopsis* Traver, 1958:491. Type-species: *Tricorythopsis artigas* Traver, original designation. Leptohyphidae: Leptohyphinae.

## Subgenus **Tricorythurus** Lestage

*Tricorythurus* (*Tricorythurus*) Lestage, 1942:13.     Type-species:
*Tricorythus latus* Ulmer, monotypy. Tricorythidae: Tricorythinae.
Subgenus of *Tricorythus* Eaton.

## Genus **Tricorythus** Eaton

*Tricorythus* Eaton, 1868a:82. Type-species: *Caenis varicauda* Pictet,
original designation. Tricorythidae: Tricorythinae.

Two subgenera are recognized: *Tricorythus* (s.s.) and *Tricorythurus* Lestage.

## Genus **Triplosoba** Handlirsch

*Triplosoba* Handlirsch, 1906:312. Type-species: *Blanchardia pulchella*
Brongniart, objective synonymy. Triplosobidae. Replacement name
for *Blanchardia* Brongniart.

*Triplosoba* is known only from fossils.

## Genus *Tritogenesia* Lestage

*Tritogenesia* Lestage, 1923:111. Type-species: *Tritogenesia bibisica*
Lestage, original designation. Palingeniidae: Palingeniinae. Synonym
of *Plethogenesia* Ulmer (Demoulin, 1965b:330).

## Genus **Turfanerella** Demoulin

*Turfanerella* Demoulin, 1954d:324. Type-species: *Ephemeropsis tingi*
Ping, original designation. Ephemerellidae: Ephemerellinae: Ephemerellini: Ephemerellae.

*Turfanerella* is known only from fossils.

## Subgenus **Ulmeritoides** Traver

*Ulmeritus* (*Ulmeritoides*) Traver, 1959:8.   Type-species: *Ulmeritus*
*uruguayensis* Traver, original designation. Leptophlebiidae: Atalophlebiinae. Subgenus of *Ulmeritus* Traver.

## Genus **Ulmeritus** Traver

*Ulmeritus* Traver, 1956:2. Type-species: *Ulmeritus carbonelli* Traver,
original designation. Leptophlebiidae: Atalophlebiinae.

Three subgenera are recognized: *Ulmeritus* (s.s.), *Pseudulmeritus* Traver, and
*Ulmeritoides* Traver.

## Genus **Ulmerophlebia** Demoulin
*Ulmerophlebia* Demoulin, 1955f:228.    Type-species: *Euphyurus mjobergi* Ulmer, original designation.    Leptophlebiidae: Atalophlebiinae.

### Subgenus **Unirachella** Allen
*Drunella (Unirachella)* Allen, 1980:80.    Type-species: *Ephemerella tuberculata* Morgan, original designation.    Ephemerellidae: Ephemerellinae: Ephemerellini: Ephemerellae.    Subgenus of *Drunella* Needham.

### Subgenus **Uracanthella** Belov
*Uracanthella* Belov, 1979:577.    Type-species: *Ephemerella lenoki* Tshernova, original designation.    Ephemerellidae: Ephemerellinae: Ephemerellini: Ephemerellae.    Subgenus of *Ephemerella* Walsh.

### Subgenus *Vernobaetis* Kazlauskas
*Baetis (Vernobaetis)* Kazlauskas, 1972:338. Not available, Art. 13b, no type designated. Baetidae.    Subgenus of *Baetis* Leach.

## Genus **Vietnamella** Tshernova
*Vietnamella* Tshernova, 1972:609.    Type-species: *Vietnamella thani* Tshernova, original designation.    Ephemerellidae: Ephemerellinae: Ephemerellini: Vietnamellae.
*Vietnamella* has been treated as a subgenus of *Cincticostella* Allen.

### Subgenus **Yaruma** Savage & Peters
*Miroculis (Yaruma)* Savage & Peters, 1983:546.    Type-species: *Miroculis wandae* Savage & Peters, original designation.    Leptophlebiidae: Atalophlebiinae.    Subgenus of *Miroculis* Edmunds.

## Genus **Xenophlebia** Demoulin
*Xenophlebia* Demoulin, 1968b:267.    Type-species: *Xenophlebia aenigmatica* Demoulin, original designation.    Leptophlebiidae: Atalophlebiinae.
*Xenophlebia* is known only from fossils.

## Genus *Xenophlebia* Riek
*Xenophlebia* Riek, 1976:150. Type-species: *Xenophlebia optata* Riek, original designation. Preoccupied. Synonym of *Litophlebia* Hubbard & Riek.

Originally proposed as a mayfly, *Xenophlebia* (the name has been replaced by *Litophlebia* Hubbard & Riek) belongs to the Megasecoptera.

## Genus **Zephlebia** Penniket
*Zephlebia* Penniket, 1961:8. Type-species: *Atalophlebia versicolor* Eaton, original designation. Leptophlebiidae: Atalophlebiinae.

Two subgenera are recognized: *Zephlebia* (s.s.) and *Terama* Towns.

# Literature Cited

Albarda, H. 1878. Descriptions of three new European Ephemeridae. Entomologist's Monthly Magazine 15:128-130.

Ali, S. R. 1970. Certain mayflies (Order: Ephemeroptera) of West Pakistan. Pakistan Journal of Science 22:119-124.

Allen, R. K. 1965. A review of the subfamilies of Ephemerellidae (Ephemeroptera). Journal of the Kansas Entomological Society 38:262-266.

Allen, R. K. 1966. *Haplohyphes*, a new genus of Leptohyphinae (Ephemeroptera: Tricorythidae). Journal of the Kansas Entomological Society 39:565-568.

Allen, R. K. 1971. New Asian *Ephemerella* with notes (Ephemeroptera: Ephemerellidae). Canadian Entomologist 103:512-528.

Allen, R. K. 1973. New Ephemerellidae from Madagascar and Afghanistan (Ephemeroptera). Pan-Pacific Entomologist 49:160-164.

Allen, R. K. 1974. *Neochoroterpes*, a new subgenus of *Choroterpes* Eaton from North America (Ephemeroptera: Leptophlebiidae). Canadian Entomologist 106:161-168.

Allen, R. K. 1980. Geographic distribution and reclassification of the subfamily Ephemerellinae (Ephemeroptera: Ephemerellidae). Pages 71-91 *in* J. F. Flannagan & K. E. Marshall, eds. Advances in Ephemeroptera Biology. Plenum, New York.

Allen, R. K. 1984. A new classification of the subfamily Ephemerellinae and the description of a new genus. Pan-Pacific Entomologist 60:245-247.

Allen, R. K. & G. F. Edmunds, Jr. 1963. New and little known Ephemerellidae from southern Asia, Africa and Madagascar. Pacific Insects 5:11-22.

Allen, R. K. & G. F. Edmunds, Jr. 1965. A revision of the genus *Ephemerella* (Ephemeroptera, Ephemerellidae). VIII. The subgenus *Ephemerella* in North America. Miscellaneous Publications of the Entomological Society of America 4:242-282.

Allen, R. K. & G. F. Edmunds, Jr. 1976. *Hyrtanella*: a new genus of Ephemerellidae from Malaysia (Ephemeroptera). Pan-Pacific Entomologist 52:133-137.

Allen, R. K. & C. M. Murvosh. 1987. Mayflies (Ephemeroptera: Tricorythidae) of the southwestern United States and northern Mexico. Annals of the Entomological Society of America 80:35-40.

Aro, J. E. 1910. Piirteitä päiväkorennoisten (Ephemeridae) elämäntavoista ja kehityksestä. Viipurin Suomalaisen Realilyseon Vuosikertomus 1909-1910. 32 pp.

Bajkova, O. Ya. 1975. [New genus of mayfly from Primorya (Ephemeroptera: Heptageniidae)] (in Russian). Izvestia Sibirskogo Otdelenia Akademii Nauk. SSSR 1(5):54-57.

Balthasar, V. 1937. Arthropleidae, eine neue Familie der Ephemeropteren. Zoologischer Anzeiger 120:204-230.

Banks, N. 1900. New genera and species of Nearctic Neuropteroid insects. Transactions of the American Entomological Society 26:239-259.

Barnard, K. H. 1932. South African May-flies (Ephemeroptera). Transactions of the Royal Society of South Africa 20:201-259.

Barnard, K. H. 1940. Additional records, and descriptions of new species, of South African alder-flies (Megaloptera), May-flies (Ephemeroptera), caddis-flies (Trichoptera), stone-flies (Perlaria), and dragon-flies (Odonata). Annals of the South African Museum 32:609-661.

Bednarik, A. F. 1979. Subgeneric classification of Stenonema (Ephemeroptera: Heptageniidae). Journal of the Georgia Entomological Society 14:190-191.

Belov, V. V. 1979. [A new genus of mayflies (Ephemeroptera, Ephemerellidae) for the fauna of the USSR] (in Russian). Doklady Akademii Nauk Ukraniskoi SSR, Ser. B, 1979: 577-580.

Bengtsson, S. 1904. Reseberättelse afgifven af Regnellske stipendiaten docenten Simon Bengtsson för en zoologisk resa til Umeå Lappmark 1903. Kungliga Svenska Vetenskaps-Akademiens Årsbok 1904:117-131.

Bengtsson, S. 1908. Berättelse öfver en resa i entomologiskt syfte till mellersta Sverige sommaren 1907. Kunglinga Svenska Vetenskaps-Akademiens Årsbok 6:237-246.

Bengtsson, S. 1909. Beiträge zur Kenntnis der paläarktischen Ephemeriden. Lunds Universitets Årsbok, N.F., Afd. 2, 5(4):1-19.

Bengtsson, S. 1912. Neue Ephemeriden aus Schweden. Entomologisk Tidskrift 33:107-117.

Bengtsson, S. 1913. Undersökningar öfver äggen hos Ephemeriderna. Entomologisk Tidskrift 34:271-320, pl. 1-3.

Bengtsson, S. 1914. Bemerkungen über die nordischen Arten der Gattung Cloëon Leach. Entomologisk Tidskrift 35:210-220.

Bengtsson, S. 1915. Eine Namensänderung. Entomologisk Tidskrift 36:34.

Bengtsson, S. 1917. Weitere Beiträge zur Kenntnis der nordischen Eintagsfliegen. Entomologisk Tidskrift 38:174-194.

Bengtsson, S. 1930. Kritische Bemerkungen über einige nordische Ephemeropteren, nebst Beschreibung neuer Larven. Lunds Universitets Årsskrift, N.F., (2)26(3):1-27.

Blair, K. G. 1929. Two new British mayflies (Ephemeroptera). Entomologist's Monthly Magazine 65:253-255.

Bogoesco, C. D. 1947. Un genre nouveau d'Ephéméroptère trouvè en Roumanie. Bulletin de la Section des Sciences de l'Academie Roumainie 29:602-606.

Braasch, D. & T. Soldán. 1980. *Centroptella* n. gen., eine neue Gattung der Eintagfliegen aus China (Baëtidae, Ephemeroptera). Reichenbachia 18:123-127.

Braasch, D. & T. Soldán. 1985. Beitrag zur Kenntnis der Gattung *Thalerosphyrus* Eaton, 1881 im Hinblick auf die Gattung *Ecdyonuroides* Thanh, 1967 (Ephemeroptera, Heptageniidae). Reichenbachia 22:201-206.

Braasch, D. & T. Soldán. 1986a. Die Heptageniidae des Gombak River in Maylaysia (Ephemeroptera). Reichenbachia 23:41-52.

Braasch, D. & T. Soldán. 1986b. *Asionurus* n. gen., eine neue Gattung der Heptageniidae von Vietnam (Ephemeroptera). Reichenbachia 23:155-159.

Braasch, D. & T. Soldán. 1988. *Trichogenia* gen. n., eine neue Gattung der Eintagsfliegen aus Vietnam (Insecta, Ephemeroptera, Heptageniidae). Reichenbachia 25:119-124.

Brauer, F., J. Redtenbacher, and L. Ganglbauer. 1889. Fossile Insekten aus der Juraformation Ost-Sibiriens. Mémoires de l'Academie Impériale des Sciences de St.-Petersbourg, (7)36(15):1-20, 2 pl.

Brito, I. M. 1987. Nota preliminar sobre uma nova Efêmera do Cretáceo do Ceará (Insecta Ephemeroptera). Anais do X Congresso Brasileiro de Paleontologia p:593-597.

Brongniart, C. 1893 [1894]. Recherches pour servir à l'histoire des Insectes fossiles des temps primaires. Bulletin de la Société Industriale et Minerale 7:124-615, pl. 17-53. Also published as Recherches pour servir à l'histoire des Insectes fossiles des temps primaires précédées d'une étude sur la nervation des ailes des Insectes. Thèses présentées a la Faculté des Sciences de Paris pour obtenir le Grade de Docteur Es-Sciences Naturelles. 493 pp. (All page citations in this catalog refer to the Thèses, since this is the only form of this work usually available.)

Buldovsky, A. T. 1935. [On new forms of the family Palingeniidae (Ephemeroptera) of the Far Eastern Region of the USSR] (in Russian). Izvestia Akademii Nauk SSSR 1935:831-836.

Burks, B. D. 1953. The mayflies, or Ephemeroptera, of Illinois. Bulletin of the Illinois Natural History Survey 26(1):1-216.

Burmeister, H. 1839. Ephemerina. Handbuch der Entomologie 2(2):788-804, 1016. Berlin.

Campbell, I. C. & P. J. Suter. 1988. Three new genera, a new subgenus and a new species of Leptophlebiidae (Ephemeroptera) from Australia. Journal of the Australian Entomological Society 27:259-273.

Campion, H. 1923. On the use of the generic name *Brachycercus* in Plectoptera and Orthoptera. Annals and Magazine of Natural History (9)11:515-518.

Carpenter, F. M. 1938. Two Carboniferous insects from the vicinity of Mazon Creek, Illinois. American Journal of Science (5)36:445-452.

Carpenter, F. M. 1980. Lower Permian insects from Oklahoma. Part 2. Orders Ephemeroptera and Palaeodictyoptera. Psyche 86:261-290.

Clemens, W. A. 1915. Mayflies of the *Siphlonurus* group. Canadian Entomologist 47:245-260, pl. 9-11.

Cockerell, T. D. A. 1923. A new genus of mayflies from the Miocene of Florissant, Colorado. Psyche 30:170-172.

Cockerell, T. D. A. 1924. Fossils in the Ondai Sair Formation, Mongolia. Bulletin of the American Museum of Natural History 51:129-144.

Crass, R. S. 1947a. Mayflies (Ephemeroptera) collected by J. Omer-Cooper in the Eastern Cape Province, South Africa, with a description of a new genus and species (*Notonurus cooperi*). Proceedings of the Royal Entomological Society of London 17:124-128.

Crass, R. S. 1947b. The May-flies (Ephemeroptera) of Natal and the Eastern Cape. Annals of the Natal Museum 11:37-110.

Curtis, J. 1834. Descriptions of some nondescript British species of May-flies of anglers. London and Edinburgh Philosophical Magazine and Journal of Science (3)4:120-125, 212-218.

Dạng Ngọc Thanh. 1967. [New genera, new species of the invertebrate fauna of fresh and brackish waters of North Vietnam] (in Vietnamese). Sinh Vật Dịa Học 6:155-165.

Day, W. C. 1953. A new mayfly genus from California (Ephemeroptera). Pan-Pacific Entomologist 29:19-24.

Day, W. C. 1955. New genera of mayflies from California (Ephemeroptera). Pan-Pacific Entomologist 31:121-137.

Dean, J. C. 1987. Two new genera of Leptophlebiidae (Insecta: Ephemeroptera) from southwestern Australia. Memoirs of the Museum of Victoria 48:91-100.

Dean, J. C. 1988. Description of a new genus of leptophlebiid mayfly from Australia (Ephemeroptera: Leptophlebiidae: Atalophlebiinae). Proceedings of the Royal Society of Victoria 100:39-45.

Demoulin, G. 1951. A propos de *Metretopus goetghebueri* Lestage, 1938, et des Metretopodidae (Insectes Ephéméroptères). Bulletin de l'Institut Royal des Sciences Naturelles de Belgique 27(49):1-20.

Demoulin, G. 1952a. Sur deux Palingeniidae (Insectes Ephéméroptères) mal connus. Bulletin de l'Institut Royal des Sciences Naturelles de Belgique 28(13):1-11.

Demoulin, G. 1952b. Essai de nouvelle clé pour le détermination des Oligoneuriidae (Insectes Ephéméroptères). Bulletin de l'Institut Royal des Sciences Naturelles de Belgique 28(43):1-4.

Demoulin, G. 1952c. Contribution à l'étude des Ephoronidae Euthyplociinae (Insectes Ephéméroptères). Bulletin de l'Institut Royal des Sciences Naturelles de Belgique 28(45):1-22.

Demoulin, G. 1953. Les Chromarcyinae subfam nov., Ephémeroptères Oligoneuriides orientaux. Bulletin de l'Institut Royal des Sciences Naturelles de Belgique 39(17):1-12.

Demoulin, G. 1954a. Description préliminaire d'un type larvaire nouveau d'Ephéméroptères Tricorythidae du Congo Belge. Bulletin de l'Institut Royal des Sciences Naturelles de Belgique 30(6):1-4.

Demoulin, G. 1954b. Une question de nomenclature: *Prosopistoma foliaceum* (Fourcroy) ou *Binoculus pennigerus* (Müller)? Bulletin et Annales de la Société Entomologique de Belgique 90:99-103.

Demoulin, G. 1954c. Les Ephéméroptères Leptophlebiidae de Borneo. Bulletin et Annales de la Société Entomologique de Belgique 90:123-131.

Demoulin, G. 1954d. Les Ephéméroptères jurassiques du Sinkiang. Bulletin et Annales de la Société Entomologique de Belgique 90:322-326.

Demoulin, G. 1954e. Essai sur quelques Ephéméroptères fossiles adultes. Volume Jubilaire Victor van Straelen (Bruxelles) 1:549-574.

Demoulin, G. 1955a. Une mission biologique belge au Brésil. Éphéméroptères. Bulletin de l'Institut Royal des Sciences Naturelles de Belgique 31(20):1-32.

Demoulin, G. 1955b. Ephéméroptères nouveaux ou rares du Chili. II. Bulletin de l'Institut Royal des Sciences Naturelles de Belgique 31(58):1-16.

Demoulin, G. 1955c. Ephéméroptères nouveaux ou rares du Chili. III. Bulletin de l'Institut Royal des Sciences Naturelles de Belgique 31(73):1-30.

Demoulin, G. 1955d. *Nathanella* gen. nov., Leptophlebiidae diptère de l'Inde (Ephemeroptera). Bulletin de l'Institut Royal des Sciences Naturelles de Belgique 31(77):1-4.

Demoulin, G. 1955e. *Melanemerella brasiliana* Ulmer, Ephémérellide ou Tricorythide? (Ephemeroptera). Bulletin et Annales de la Société Royale Entomologique de Belgique 91:214-216.

Demoulin, G. 1955f. Note sur deux nouveaux genres de Leptophlebiidae d'Australie (Ephemeroptera). Bulletin et Annales de la Société Royale Entomologique de Belgique 91:227-229.

Demoulin, G. 1955g. Sur une larve siphlonuridienne d'Ephémère fossile du Brésil. Bulletin et Annales de la Société Royale d'Entomologie de Belgique 91:271.

Demoulin, G. 1955h. Revision de quelques Ephéméroptères décrits du Congo belge par L. Navás. I. Bulletin et Annales de la Société Royale Entomologique de Belgique 91:281-290.

Demoulin, G. 1955i. *Afromera* gen. nov., Ephemeridae de la faune éthiopienne (Ephemeroptera). Bulletin et Annales de la Société Royale Entomologique de Belgique 91:291-295.

Demoulin, G. 1955j. Quelques remarques sur les composantes de la famille Ametropodidae (Ephemeroptera). Bulletin et Annales de la Société Royale Entomologique de Belgique 91:342-346.

Demoulin, G. 1956a. *Electrogenia dewalschei* n. gen. n. sp., Ephéméroptère fossile de l'ambre. Bulletin et Annales de la Société Royale d'Entomologie de Belgique 92:95-100.

Demoulin, G. 1956b. Quelques Ephéméroptères du Kivu. Bulletin et Annales de la Société Royale Entomologique de Belgique 92:277-284.

Demoulin, G. 1957. Remarques critiques sur la position systématique des *Ichthybotus* Eaton, Ephéméroptères de Nouvelle-Zélande. Bulletin et Annales de la Société Royale d'Entomologie de Belgique 93:335-337.

Demoulin, G. 1959. Une curieuse larve d'Ephéméroptères de l'Angola portugais. Bulletin et Annales de la Société Royale Entomologique de Belgique 95:249-252.

Demoulin, G. 1961. A propos de données recentes sur la "*Caenis*" *maxima* Joly (Ephemeroptera). Bulletin et Annales de la Société Royale Entomologique de Belgique 97:63-68.

Demoulin, G. 1962. A propos des données recentes sur "*Leucorhoenanthus*" *macedonicus* (Ulmer) (Ephemeroptera). Bulletin et Annales de la Société Royale d'Entomologie de Belgique 98:368-370.

Demoulin, G. 1964a. Ephemeroptera. Parc National de l'Upemba - Mission G. F. de Witte 68(2):13-27.

Demoulin, G. 1964b. Mission H. G. Amsel en Afghanistan (1956). Ephemeroptera. Bulletin et Annales de la Société Royale Entomologique de Belgique 100:351-363.

Demoulin, G. 1965a. Contribution à la connaissance des Ephéméroptères de l'ambre oligocène de la Baltique. Entomologiske Meddelelser 34:143-153.

Demoulin, G. 1965b. Contribution à l'étude des Palingeniidae (Insectes, Ephemeroptera). Nova Guinea, Zoology 33:305-344.

Demoulin, G. 1966a. Contribution à l'étude des Ephéméroptères du Surinam. Bulletin de l'Institute Royal des Sciences Naturales de Belgique 42(37):1-22.

Demoulin, G. 1966b. Contribution à l'étude des Euthyplociidae (Ephemeroptera). IV. Un nouveau genre de Madagascar. Annales de la Société Entomologique de France (N.S.) 2:941-949.

Demoulin, G. 1967. Redescription de l'holotype ♀ imago de *Chromarcys magnifica* Navás et discussion des affinities phyletiques du genre *Chromarcys* Navás (Ephemeroptera, Chromarcyinae). Bulletin de l'Institut Royal des Sciences Naturelles de Belgique 43(31):1-10.

Demoulin, G. 1968a. Remarques sur la position systematique de deux Ephéméroptères du Jurassique inferieur de Siberie orientale. Bulletin de l'Institut Royal des Sciences Naturelles de Belgique 44(18):1-8.

Demoulin, G. 1968b. Deuxième contribution à la connaissance des Ephéméroptères de l'ambre oligocène de la Baltique. Deutsche Entomologische Zeitschrift, N.F., 15:233-276.

Demoulin, G. 1970a. Troisième contribution à la connaissance des Ephéméroptères de l'ambre oligocène de la Baltique. Bulletin de l'Institut Royal des Sciences Naturelles de Belgique 46(2):1-11.

Demoulin, G. 1970b. Remarques critiques sur des larves "Ephemeromorphes" du Permien. Bulletin de l'Institut Royal des Sciences Naturelles de Belgique 46(3):1-10.

Demoulin, G. 1970c. Contribution à l'étude morphologique systematique et phylogenique des Ephéméroptères jurassiques d'Europe centrale. V. Hexagenitidae = Paedephemeridae (syn. nov.). Bulletin de l'Institut Royal des Sciences Naturelles de Belgique 46(4):1-8.

Demoulin, G. 1970d. Ephemeroptera des faunes éthiopienne et malgache. South African Animal Life 14:24-170.

Demoulin, G. 1973. Ephéméroptères de Madagascar. III. Bulletin de l'Institut Royal Entomolgique de Belgique 49(7):1-20.

Demoulin, G. 1974. Remarques critiques sur les Acanthametropodinae et sur certaines formes affines (Ephemeroptera Siphlonuridae). Bulletin de l'Institut Royal des Sciences Naturelles de Belgique 50(2):1-5.

Dominguez, E. 1988. *Ecuaphlebia*: A new genus of Atalophlebiinae (Ephemeroptera: Leptophlebiidae) from Ecuador. Aquatic Insects 10:227-235.

Dominguez, E. & R. W. Flowers. 1989. A revision of *Hermanella* and related genera (Ephemeroptera: Leptophlebiidae: Atalophlebiinae) from Subtropical South America. Annals of the Entomological Society of America 82:555-573.

Eaton, A. E. 1866. Notes on some species of the orthopterous genus *Cloeon*, Leach (as limited by M. Pictet). Annals and Magazine of Natural History (3)18:145-148.

Eaton, A. E. 1868a. An outline of a re-arrangement of the genera of Ephemeridae. Entomologist's Monthly Magazine 5:82-91.

Eaton, A. E. 1868b. Remarks upon the homologies of the ovipositor. Transactions of the Entomological Society 1868:141-144.

Eaton, A. E. 1869. On *Centroptilum*, a new genus of the Ephemeridae. Entomologist's Monthly Magazine 6:131-132.

Eaton, A. E. 1871. A monograph on the Ephemeridae. Transactions of the Entomological Society of London 1871:1-164, pl. 1-6.

Eaton, A. E. 1881a. An announcement of new genera of the Ephemeridae. Entomologist's Monthly Magazine 17:191-197.

Eaton, A. E. 1881b. An announcement of new genera of the Ephemeridae. Entomologist's Monthly Magazine 18:21-27.

Eaton, A. E. 1882. An announcement of new genera of the Ephemeridae. Entomologist's Monthly Magazine 18:207-208.

Eaton, A. E. 1883-1888. A revisional monograph of Recent Ephemeridae or mayflies. Transactions of the Linnean Society of London, Zoology (2)3:1-352, pl. 1-65.

Eaton, A. E. 1899. An annotated list of the Ephemeridae of New Zealand. Transactions of the Entomological Society of London 1899:285-293, 1 pl.

Eaton, A. E. 1901. Ephemeridae collected by Herr E. Strand in South and Arctic Norway. Entomologist's Monthly Magazine (2)12:252-255.

Edmunds, G. F. Jr. 1948. A new genus of mayflies from western North America (Leptophlebiinae). Proceedings of the Entomological Society of Washington 61:141-146.

Edmunds, G. F., Jr. 1953. Taxonomic notes on the genus *Adenophlebiodes* Ulmer (Ephemeroptera: Leptophlebiidae). Revue Zoologique et Botanique Africaine 48:79-80.

Edmunds, G. F., Jr. 1957. The systematic relationships of the Paleantarctic Siphlonuridae (including Isonychiidae) (Ephemeroptera). Proceedings of the Entomological Society of Washington 59:245-246.

Edmunds, G. F., Jr. 1959. Subgeneric groups within the mayfly genus *Ephemerella* (Ephemeroptera: Ephemerellidae). Annals of the Entomological Society of America 52:543-547.

Edmunds, G. F., Jr. 1960. Two generic synonyms in the Siphlonuridae (Ephemeroptera). Bulletin of the Brooklyn Entomological Society 55:24.

Edmunds, G. F., Jr. 1962a. The type localities of the Ephemeroptera of North America north of Mexico. University of Utah Biological Series 12(5):1-39.

Edmunds, G. F., Jr. 1962b. The principles applied in determining the hierarchic level of the higher categories of Ephemeroptera. Systematic Zoology 11:22-31.

Edmunds, G. F., Jr. 1963. A new genus and species of mayfly from Peru (Ephemeroptera: Leptophlebiidae). Pan-Pacific Entomologist 39:34-36.

Edmunds, G. F., Jr. 1971. A new name for a subgeneric homonym in *Ephemerella* (Ephemeroptera: Ephemerellidae). Proceedings of the Entomological Society of Washington 73:152.

Edmunds, G. F., Jr. 1974. Some taxonomic changes in Baetidae (Ephemeroptera). Proceedings of the Entomological Society of Washington 76:289.

Edmunds, G. F., Jr., & R. K. Allen. 1957. A checklist of the Ephemeroptera of North America north of Mexico. Annals of the Entomological Society of America 50:317-324.

Edmunds, G. F. Jr., R. K. Allen & W. L. Peters. 1963. An annotated key to the nymphs of the families and subfamilies of mayflies (Ephemeroptera). University of Utah Biological Series 13(1):1-49.

Edmunds, G. F., Jr., & S. L. Jensen. 1974. A new genus and subfamily of North American Heptageniidae (Ephemeroptera). Proceedings of the Entomological Society of Washington 76:495-497.

Edmunds, G. F., Jr., S. L. Jensen & L. Berner. 1976. The mayflies of North and Central America. University of Minnesota Press, Minneapolis. 330p.

Edmunds, G. F., Jr., & R. W. Koss. 1972. A review of the Acanthametropodinae with a description of a new genus (Ephemeroptera: Siphlonuridae). Pan-Pacific Entomologist 48:136-144.

Edmunds, G. F., Jr., & J. R. Traver. 1954. An outline of a reclassification of the Ephemeroptera. Proceedings of the Entomological Society of Washington 56:236-240.

Edmunds, G. F., Jr., & J. R. Traver. 1959. The classification of the Ephemeroptera. I. Ephemeroidea: Behningiidae. Annals of the Entomological Society of America 52:43-51.

d'Eichwald, E. 1864. Sur un terrain jurassique à poissons et Insectes d'eau douce de la Sibérie orientale. Bulletin de la Société Géologique de France (2)21:19-25.

Esben Petersen, [T.] 1909. New Ephemeridae from Denmark, Arctic Norway and the Argentine Republic. Deutsche Entomologische Zeitschrift 1909:551-556.

Flowers, R. W. 1980. *Atopophlebia fortunensis*, a new genus and species from Panamá (Leptophlebiidae: Ephemeroptera). Florida Entomologist 63:162-165.

Flowers, R. W. 1980. Two new genera of Nearctic Heptageniidae (Ephemeroptera). Florida Entomologist 63:296-307.

Flowers, R. W. 1985. *Guajirolus*, a new genus of Neotropical Baetidae (Ephemeroptera). Studies on Neotropical Fauna and Environment 20:27-31.

Geoffroy, E. L. 1762. Histoire abregée des insectes qui se trouvent aux environs de Paris, Vol. 2. Paris, Durand.

Gillies, M. T. 1951. Further notes on Ephemeroptera from India and South East Asia. Proceedings of the Royal Entomological Society of London (B)20:121-130.

Gillies, M. T. 1957. New records and species of *Euthraulus* Barnard (Ephemeroptera) from East Africa and the Oriental Region. Proceedings of the Royal Entomological Society of London B26:43-48.

Gillies, M. T. 1960. A new genus of Tricorythidae (Ephemeroptera) from East Africa. Proceedings of the Royal Entomological Society of London (B)29:35-40.

Gillies, M. T. 1977. A new genus of Caenidae (Ephemeroptera) from East Africa. Journal of Natural History 11:451-455.

Gillies, M. T. 1980. The African Euthyplociidae (Ephemeroptera) (Exeuthyplociinae subfam. n.). Aquatic Insects 2:217-224.

Gillies, M. T. 1982. A second large-eyed genus of Caenidae (Ephemeroptera) from Africa. Journal of Natural History 16:15-22.

Gillies, M. T. 1984. On the synonymy of *Notonurus* Crass with *Compsoneuriella* Ulmer (Heptageniidae). Pages 21-25 *in* V. Landa, T. Soldán & M. Tonner, eds. Proceedings of the IV International Conference on Ephemeroptera.

Gillies, M. T. 1985. A preliminary account of the East African species of *Cloeon* Leach and *Rhithrocloeon* gen. n. (Ephemeroptera). Aquatic Insects 7:1-17.

Gillies, M. T. & J.-M. Elouard. 1990. The mayfly-mussel association, a new example from the River Niger basin. Pages 289-297 *in* I. C. Campbell, ed. Mayflies and Stoneflies: Life Histories and Biology, Kluwer Academic Publishers, Dordrecht.

Gros, A. J. & J. A. Lestage. 1927. Contribution à l'étude des larves des Ephéméroptères. Série IV. Le groupe Euthyplocien. Annales de Biologie Lacustre 15:119-162, 1 pl.

Hagen, H. 1868. On *Lachlania abnormis*, a new genus and species from Cuba belonging to the Ephemerina. Proceedings of the Boston Society of Natural History 11:372-375.

Handlirsch, A. 1904. Über einige Insektenreste aus der Permformation Russlands. Zapiski Imperatorskoi Akademii Nauk po Fiziko-Matematicheskomu Otdbleniyu (8)16(5):1-8, 1 tab.

Handlirsch, A. 1906-1908. Die fossilen Insekten und die Phylogenie der rezenten Formen. Ein Handbuch für Paläontologen und Zoologen. Verlag von Wilhelm Engelmann, Leipzig. 1430p.

Handlirsch, A. 1911. New Paleozoic insects from the vicinity of Mazon Creek, Illinois. American Journal of Science (4)31:297-326, 353-377.

Handlirsch, A. 1918. Fossile Ephemeridenlarven aus dem Buntsandstein der Vogesen. Verhandlungen der Kaiserlich-Koniglichen Zoologisch-Botanischen Gesellschaft in Wien 68:112-114.

Harker, J. E. 1954. The Ephemeroptera of Eastern Australia. Transactions of the Royal Entomological Society of London 105:241-268.

Harker, J. E. 1957. Some new Australian Ephemeroptera. Proceedings of the Royal Entomological Society of London (B)26:63-78.

Haupt, H. 1956. Beitrag zur Kenntnis der eozänen Arthropodenfauna des Geiseltales. Nova Acta Leopoldina, N.F., 18(128):1-90.

Hong, Y.-C. 1979. On Eocene *Philolimnias* gen. nov. (Ephemeroptera, Insecta) in amber from Fushun Coalfield, Liaoning Province. Scientia Sinica 22:331-339, 2 pl.

Hsu, Y.-C. 1937-8. The mayflies of China. Peking Natural History Bulletin (1937)11:129-148, 287-296, 433-440; (1938)12:53-56, 125-126, 221-224.

Hubbard, M. D. 1977. The validity of the generic name *Parameletus* Bengtsson (Ephemeroptera: Siphlonuridae). Proceedings of the Entomological Society of Washington 79:409-410.

Hubbard, M. D. 1979a. Designation of type species for 2 genera of Siphlonuridae (Ephemeroptera: Siphlonuridae). Florida Entomologist 62:412.

Hubbard, M. D. 1979b. Genera and subgenera of Recent Ephemeroptera. Eatonia Supplement no. 2:1-8.

Hubbard, M. D. 1981. Type-species designation for the Jurassic mayfly genus *Mesephemera* (Ephemeroptera: Mesephemeridae). Great Lakes Entomologist 14:69.

Hubbard, M. D. 1982a. Catálogo abreviado de Ephemeroptera da América do Sul. Papéis Avulsos de Zoologia 34:257-282.

Hubbard, M. D. 1982b. Catalog of the Ephemeroptera: Family-group taxa. Aquatic Insects 4:49-53.

Hubbard, M. D. 1984. A revision of the genus *Povilla* (Ephemeroptera: Polymitarcyidae). Aquatic Insects 6:17-35.

Hubbard, M. D. 1985. The nomenclature of *Murphyella* and *Dictyosiphlon* (Ephemeroptera: Siphlonuridae: Coloburiscinae). Revista Chilena de Entomologia 12:11-13.

Hubbard, M. D. 1986. A catalog of the mayflies (Ephemeroptera) of Hong Kong. Insecta Mundi 1: 247-254.

Hubbard, M. D. 1987. Ephemeroptera. Fossilium Catalogus. I: Animalia. Pars 129. iii + 99 pp.

Hubbard, M. D. 1988. *Promatsumura*, replacement name for *Procloeon* Matsumura, 1931 (Ephemeroptera: Baetidae) with designation of type-species. Insecta Mundi 2:240.

Hubbard, M. D. 1989. *Matsumuracloeon*, a replacement name for *Pseudocloeon* Matsumura, 1931 (Ephemeroptera: Baetidae). Florida Entomologist 72:388.

Hubbard, M. D. & J. Kukalová-Peck. 1980. Permian mayfly nymphs: new taxa and systematic characters. Pages 19-31 *in* J. F. Flannagan & K. E. Marshall, eds. Advances in Ephemeroptera Biology, Plenum Press, New York.

Hubbard, M. D. & M. L. Pescador. 1978. A catalog of the Ephemeroptera of the Philippines. Pacific Insects 19:91-99.

Hubbard, M. D. & W. L. Peters. 1976. The number of genera and species of mayflies (Ephemeroptera). Entomological News 87:245.

Hubbard, M. D. & W. L. Peters. 1978. A catalogue of the Ephemeroptera of the Indian Subregion. Oriental Insects Supplement 9:1-43.

Hubbard, M. D. & E. F. Riek. 1978. New name for a Triassic mayfly from South Africa (Ephemeroptera). Psyche 83:260-261.

Hubbard, M. D. & H. M. Savage. 1981. The fossil Leptophlebiidae (Ephemeroptera): A systematic and phylogenetic review. Journal of Paleontology 55:810-813.

Ide, F. P. 1935. Life history notes on *Ephoron*, *Potamanthus*, *Leptophlebia* and *Blasturus* with descriptions (Ephemeroptera). Canadian Entomologist 67:113-125.

Imanishi, K. 1935. Mayflies from Japanese torrents. V. Notes of the genera *Cinygma* and *Heptagenia*. Annotationes Zoologicae Japoneses 15:213-223.

International Commission on Zoological Nomenclature. 1954. Opinion 228. Opinions and Declarations Rendered by the International Commission on Zoological Nomenclature 4(18):209-220.

International Commission on Zoological Nomenclature. 1966. Opinion 787. Bulletin of Zoological Nomenclature 23:209-210.

Jacob, U. 1974. Zur Kenntnis zweier *Oxycypha*-Arten Hermann Burmeisters (Ephemeroptera, Caenidae). Reichenbachia 15:93-97.

Jacob, U. 1984. Larvale Oberflächenskulpturen bei Ephemeropteren und ihr Wert für Taxonomie und Systematik. Pages 181-191 *in* V. Landa, T. Soldán & M. Tonner, eds. Proceedings of the IVth International Conference on Ephemeroptera.

Jacob, U. & A. Glazaczow. 1986. *Pseudocentroptiloides*, a new baetid genus of Palaearctic and Oriental disbtribution [sic] (Ephemeroptera). Aquatic Insects 8:197-206.

Jacobson, G. G. & V. L. Bianki. 1905. [Orthoptera and Pseudoneuroptera of the Russian Empire and bordering countries.] (in Russian). Izdanie A. F. Devriena, St. Petersburg. 952p.

Jell, P. A. & P. M. Duncan. 1986. Invertebrates, mainly insects, from the freshwater, Lower Cretaceous, Koonwarra Fossil Bed (Korumburra Group), South Gippsland, Victoria. Pages 111-205 *in* P. A. Jell & J. Roberts, eds. Plants and Invertebrates from the Lower Cretaceous Koonwarra Fossil Bed, South Gippsland, Victoria, Memoirs of the Association of Australasian Palaeontologists, vol. 3.

Jensen, S. L. 1974. A new genus of mayflies from North America (Ephemeroptera: Heptageniidae). Proceedings of the Entomological Society of Washington 76:225-228.

Joly, N. & E. Joly. 1872. Etudes sur le prétendu crustacé. Mémoires de l'Academie des Sciences, Inscriptions et Belles-Lettres de Toulouse (7)4:437-438.

Joly, N. & E. Joly. 1877. Contributions à l'histoire naturelle et à l'anatomie des Ephémerines. Revue de l'Academie des Sciences 5:305-330.

Kazlauskas, R. S. 1972. Neues über das System der Eintagsfliegen der Familie Baetidae (Ephemeroptera). Proceedings of the 13th International Congress of Entomology, Moscow, 1968, 3:337-338.

Keffermüller, M. 1960. Badania nad fauną jętek (Ephemeroptera) Wielkopolski. Prace Komisji Biologicznej Poznańskie Towarzystwo Przyjaciół Nauk Wydział Matematiyczno-Przyrodniczy 19(8):411-467, pl. 1-11.

Keffermüller, M. & R. Sowa. 1984. Survey of Central European species of the genera *Centroptilum* Eaton and *Pseudocentroptilum* Bogoescu (Ephemeroptera, Baetidae). Polskie Pismo Entomologiczne 54:309-340.

Kimmins, D. E. 1937. Some new Ephemeroptera. Annals and Magazine of Natural History (10)19:430-440, pl. 11.

Kimmins, D. E. 1947. New species of Indian Ephemeroptera. Proceedings of the Royal Entomological Society of London (B)16:92-100.

Kimmins, D. E. 1949. Ephemeroptera from Nyasaland, with descriptions of new species. Annals and Magazine of Natural History (12)1:825-836.

Klapálek, F. 1905. Plecopteren und Ephemeriden aus Java, gesammelt von Prof. K. Kraepelin 1904. Mitteilungen aus dem Naturhistorischen Museum in Hamburg 22:103-107.

Klapálek, F. 1909. Ephemerida, Eintagsfliegen. Süsswasserfauna Deutschlands 8:1-32.

Kluge, N. Yu. 1980. [To the knowledge of mayflies (Ephemeroptera) of Taimyr National District] in Russian. Entomologicheskoe Obozrenie 59:561-579.

Kluge, N. Yu. 1987. [Mayflies of the genus *Heptagenia* Walsh (Ephemeroptera, Heptageniidae) of the fauna of the USSR] (in Russian). Entomologichskoe Obozrenie 66:302-320.

Kondratieff, B. C. & J. R. Voshell, Jr. 1983. Subgeneric and species-group classification of the mayfly genus *Isonychia* in North America (Ephemeroptera: Oligoneuriidae). Proceedings of the Entomological Society of Washington 85:128-138.

Kukalová-Peck, J. 1985. Ephemeroid wing venation based upon new gigantic Carboniferous mayflies and basic morphology, phylogeny, and metamorphosis of pterygote insects (Insecta, Ephemerida). Canadian Journal of Zoology 63:933-955.

Lameere, A. 1917a. Etude sur l'evolution des Ephémères. Bulletin de la Société Zoologique de France 42:41-59, 61-81.

Landa, V. 1973. A contribution to the evolution of the order Ephemeroptera based on comparative anatomy. Pages 155-159 *in* W. L. Peters & J. G. Peters (eds.), Proceedings of the First International Conference on Ephemeroptera. E. J. Brill, Leiden.

Landa, V. & T. Soldán. 1985. Phylogeny and higher classification of the order Ephemeroptera: A discussion from the comparative anatomical point of view. Studie Československá Akademie Věd 4:1-121.

Latreille, P. A. 1810. Considérations Générales sur l'Ordre Naturel des Animaux Composant les Classes des Crustacés, des Arachnides, et des Insects. F. Schoell, Paris. 444 pp.

Latreille, [P. A.] 1833. Description d'un nouveau genre de Crustacés. Annales du Musee d'Histoire Naturelle, Paris, 3e Sér., 2:23-34.

Leach, W. E. 1815. Entomology. Brewster's Edinburgh Encyclopaedia 9:57-172.

Lehmkuhl, D. M. 1979. A new genus and species of Heptageniidae (Ephemeroptera) from western Canada. Canadian Entomologist 111:859-862.

Lestage, J. A. 1917. Contribution à l'étude des larves de Ephéméres paléarctiques. Annales de Biologie Lacustre 8:215-458.

Lestage, J.-A. 1918. Les Ephémères d'Afrique (Notes critiques sur les espèces connues). Revue Zoologique Africaine 6:65-114.

Lestage, J.-A. 1921. Les Ephémères indo-chinoises. Annales de la
    Société Entomologique de Belgique 61:211-222.
Lestage, J.-A. 1922. Notes sur le genre *Nirvius* Navás (= *Ephemera* L.)
    [Ephemeroptera]. Bulletin de la Société Entomologique de France
    16:253-254.
Lestage, J.-A. 1923. Etude sur les Palingeniidae (Ephémères) et descrip-
    tion de deux genres nouveaux et d'une espèce nouvelle de la
    Nouvelle Guinée. Annales de la Société Entomologique de
    Belgique 63:95-112.
Lestage, J. A. 1924a. Faune entomologique de l'Indochine Française. Les
    Ephémères de l'Indochine Française. Opuscules de l'Institute
    Scientifique de l'Indochine, no. 3:79-93.
Lestage, J. A. 1924b. Les Ephémères de l'Afrique du Sud. Catalogue
    critique & systematique des espèces connues et description de
    trois genres nouveaux et de sept espèces nouvelles. Revue
    Zoologique Africaine 12:316-352.
Lestage, J.-A. 1924c. A propos du genre *Coenis* Steph. = *Brachycercus*
    Curt. (Ephemeroptera). Annales de la Société Entomologique de
    Belgiqe 64:61-62.
Lestage, J. A. 1930a. Contribution à l'étude des larves des Éphéméro-
    ptères. V. Les larves à trachéo-branchies ventrales. Bulletin et
    Annales de la Société Entomologique de Belgique 69:433-440.
Lestage, J.-A. 1930b. Notes sur le genre *Massartella* nov. gen. de la
    famille des Leptophlebiidae (Ephemeroptera) et le génotype
    *Massartella brieni* Lest. Une Mission Biologique Belge au Brésil
    2:249-258.
Lestage, J. A. 1931a. Contribution à l'étude des larves des Ephéméro-
    ptères. VII. -Le groupe Potamanthidien. Mémories de la Société
    Entomologique de Belgique 23:73-146.
Lestage, J. A. 1931b. Contribution à l'étude des Ephéméroptères. VIII.
    Les Ephéméroptères du Chili. Bulletin et Annales de la Société
    Entomologique de Belgique 71:41-60.
Lestage, J. A. 1931c. Notes sur le premier Brachycercidien découvert
    dans la faune australienne *Tasmanocoenis tonnoiri* sp. nov.
    (Ephemeroptera) et remarques sur la famille des Brachycercidae
    Lest. Mémoires de la Société Entomologique de Belgique 23:49-
    60.
Lestage, J. A. 1935. Contribution à l'étude des Ephéméroptères. XII. Les
    composantes australiennes et néo-zélandaises du Groupe Siphlo-
    nuridien. Bulletin et Annales de la Société Entomologique de
    Belgique 75:346-358.

Lestage, J. A. 1938a. Contribution à l'étude des Éphémétoptères. XVI. Recherches critiques sur le complexe amétropo-métrétopodidien. Bulletin et Annales de la Société Entomologique de Belgique 78:155-182.

Lestage, J. A. 1938b. Contribution à l'étude des Ephéméroptères. XIX. L'inclusion des Behningeniidae parmi les Oligoneuriidae. Bulletin et Annales de la Société Entomologique de Belgique 78:315-319.

Lestage, J. A. 1939. Contribution à l'étude des Ephéméroptères. XXIII. Les Polymitarcyidae de la faune africaine et description d'un genre nouveau du Natal. Bulletin et Annales de la Société Entomologique de Belgique 79:135-138.

Lestage, J. A. 1940. Contribution à l'étude des Ephéméroptères. XXIV. Un cas de non-agnathisme chez l'adult de Paleoameletus primitivus Trav. de l'Himalaya. Bulletin et Annales de la Société Entomologique de Belgique 80:118-124.

Lestage, J. A. 1942. Contribution à l'étude des Ephéméroptères. XXV. Notes critiques sur les anciens Caenidiens d'Afrique et sur l'indépendance de l'évolution Tricorythido-caenidienne. Bulletin du Musee Royal d'Histoire Naturelle de Belgique 18(48):1-20.

Lestage, J. A. 1945. Contribution à l'étude des Ephéméroptères. XXVI. Etude critique de quelques genres de la faune éthiopienne. Bulletin et Annales de la Société Entomologique de Belgique 81:81-89.

Lin, Q.-B. 1980. [Mesozoic fossil insects of Zhejiang and Anhui] (in Chinese). Pages 211-234, pl. 1-8, in Fossils of Mesozoic Deposits of Volcanic Origin in Zhejiang and Anhui, Science Press, Peking.

Lin, Q.-B. 1986. [Early Mesozoic fossil insects from South China] (in Chinese). Palaeontologia Sinica 170 (N.S. B21):i-iii, 1-112, 20 pl.

Linnaeus, C. 1758. Systema Naturae. Ed. X. Holmae. Vol. 1

McCafferty, W. P. 1968. A new genus and species of Ephemeridae (Ephemeroptera) from Madagascar. Entomologist's Record and Journal of Variation 80:293.

McCafferty, W. P. 1971. New genus of mayflies from eastern North America (Ephemeroptera: Ephemeridae). Journal of the New York Entomological Society 79:45-51.

McCafferty, W. P. 1972. Pentageniidae: a new family of Ephemeroidea (Ephemeroptera). Journal of the Georgia Entomological Society 7:51-56.

McCafferty, W. P. 1987. New fossil mayfly in amber and its relationships among extant Ephemeridae (Ephemeroptera). Annals of the Entomological Society of America 80:472-474.

McCafferty, W. P. 1989. Characterization and relationships of the subgenera of *Isonychia* (Ephemeroptera: Oligoneuriidae). Entomological News 100:72-78.

McCafferty, W. P. & G. F. Edmunds, Jr. 1973. Subgeneric classification of *Ephemera* (Ephemeroptera: Ephemeridae). Pan-Pacific Entomologist 49:300-307.

McCafferty, W. P. & G. F. Edmunds, Jr. 1976. The larvae of the Madagascar genus *Cheirogenesia* Demoulin (Ephemeroptera: Palingeniidae). Systematic Entomology 1:189-194.

McCafferty, W. P. & M. T. Gillies. 1979. The African Ephemeridae (Ephemeroptera). Aquatic Insects 1:169-178.

McCafferty, W. P. & A. V. Provansha. 1975. Reinstatement and biosystematics of *Heterocloeon* McDunnough (Ephemeroptera: Baetidae). Journal of the Georgia Entomological Society 10:123-127.

McCafferty, W. P. & A. V. Provansha. 1985. *Amercaenis*: New Nearctic genus of Caenidae (Ephemeroptera). International Quarterly of Entomology 1:1-7.

McCafferty, W. P. & A. V. Provansha. 1988. Revisionary notes on predaceous Heptageniidae based on larval and adult associations (Ephemeroptera). Great Lakes Entomologist 21:15-17.

McDunnough, J. 1923. New Canadian Ephemeridae with notes. Canadian Entomologist 55:39-50.

McDunnough, J. 1925. New Canadian Ephemeridae with notes. III. Canadian Entomologist 57:168-176, 185-192.

McDunnough, J. 1932. Further notes on the Ephemeroptera of the north shore of the Gulf of St. Lawrence. Canadian Entomologist 64:78-81.

McDunnough, J. 1933. The nymph of *Cinygma integrum* and description of a new heptagenine genus. Canadian Entomologist 65:73-76.

McLachlan, R. 1873. *Oniscigaster wakefieldi*, a new genus and species of Ephemeridae from New Zealand. Entomologist's Monthly Magazine 10:108-110.

Malzacher, P. 1987. Eine neue Caeniden-Gattung *Afrocaenis* gen. nov. und Bemerkungen zu *Tasmanocaenis tillyardi* (Insecta: Ephemeroptera). Stuttgarter Beiträge zur Naturkunde, Serie A, Nr. 407:1-10.

Martynov, A. 1928. Permian fossil insects of North-East Europe. Trudy Geologicheskogo Muzeya Akademii Nauk SSSR 4:1-118, 19 pl.

Martynov, A. 1932. New Permian Palaeoptera with the discussion of some problems of their evolution. Trudy Paleozoologicheskogo Instituta Akademiia Nauk SSSR 1:1-44, 2 pl.

Martynov, A. 1938. [Studies on the geologic history and phylogeny of the orders of insects (Pterygota). Part I. Palaeoptera and Neoptera-Polyneoptera] (in Russian). Trudi Paleontologicheskogo Instityuta 7(4):1-148, 1 pl.

Matsumura, S. 1931. [Ephemerida] (in Japanese). 6000 Illustrated Insects of the Japanese Empire :1465-1480.

Mol, A. W. M. 1986. *Harpagobaetis gulosus* gen. nov., sp. nov., a new mayfly from Suriname (Ephemeroptera: Baetidae). Zoologische Mededelingen 60:63-70.

Mol, A. W. M. 1989. *Echinobaetis phagas* gen. nov., spec. nov., a new mayfly from Sulawesi (Ephemeroptera: Baetidae). Zoologische Mededelingen 63:61-72.

Morihara, D. K. & G. F. Edmunds, Jr. 1980. *Notobaetis*: A new genus of Baetidae (Ephemeroptera) from South America. Internationale Revue der Gesamten Hydrobiologie 65:605-610.

Motaş, C. & M. Bačesco. 1937. Le découverte en Roumanie d'une nymphe d'Ephémère appartenant au genre *Behningia* J. A. Lestage 1929. Annales Scientifiques de l'Université de Jassy, Seconde Partie, 24:25-29.

Müller-Liebenau, I. 1974. *Rheobaetis*: a new genus from Georgia (Ephemeroptera: Baetidae). Annals of the Entomological Society of America 67:555-567.

Müller-Liebenau, I. 1978. *Raptobaetopus*, eine neue carnivore Ephemeroptern-Gattung aus Malaysia (Insecta, Ephemeroptera: Baetidae). Archiv für Hydrobiologie 82:465-481.

Müller-Liebenau, I. 1980. *Jubabaetis* gen. n. and *Platybaetis* gen. n., two new genera of the family Baetidae from the Oriental Region. Pages 103-114 *in* J. F. Flannagan & K. E. Marshall, eds. Advances in Ephemeroptera Biology. Plenum, New York.

Müller-Liebenau, I. 1982. A new genus and species of Baetidae from Sri Lanka (Ceylon): *Indocloeon primum* gen. n., sp. n. (Insecta, Ephemeroptera). Aquatic Insects 4:125-129.

Müller-Liebenau, I. 1985. Baetidae from Taiwan with remarks on *Baetiella* Ueno, 1931 (Insecta, Ephemeroptera). Archiv für Hydrobiologie 104:93-110.

Müller-Liebenau, I. & W. H. Heard. 1979. *Symbiocloeon*: a new genus of Baetidae from Thailand (Insecta, Ephemeroptera). Pages 57-65, 2 pl. *in* K. Pasternak & R. Sowa, eds. Proceedings of the 2nd International Conference on Ephemeroptera. Państwowe Wydawnictwo Naukowe, Warzawa-Kraków.

Müller-Liebenau, I. & D. K. Morihara. 1982. *Indobaetis*: A new genus of Baetidae from Sri Lanka (Insecta: Ephemeroptera) with two new species. Gewässer und Abwässer 68/69:26-34.

Navás, L. 1912a. Insectos neurópteros nuevos. Verhandlungen VIII Internazionale Zoologische Kongress, Graz 1910:746-751.

Navás, L. 1912b. Quelques Nevroptères de la Sibérie méridionale-orientale. Russki Entomologicheskoe Obozrenie 12:414-422.

Navás, L. 1912c. Notes sur quelques Névroptères d'Afrique. Revue Zoologique Africaine 1:401-410.

Navás, L. 1913. Algunos órganos de la alas de los insectos. II International Congress of Entomology, Oxford, 1912, 2:178-186.

Navás, L. 1915. Neue Neuropteren. Erste Serie. Entomologische Mitteilungen 4:146-153.

Navás, L. 1918. Insectos chilenos. Boletin de la Sociedad Aragonesa de Ciencias Naturales 17:212-230.

Navás, L. 1920. Insectos Sudamericanos. 3ª Serie. Anales de la Sociedad Científica Argentina 90:52-72.

Navás, L. 1922a. Efemerópteros nuevos o poco conocidos. Boletin de la Sociedad Entomológica de España 5:54-63.

Navás, L. 1922b. Insectos de Fernando Poo. Treballs del Museu de Ciències Naturals de Barcelona 4:107-116.

Navás, L. 1924. Insectos de la América Central. Broteria 21:55-86.

Navás, L. 1930. Insectes du Congo Belge (Serie IV). Revue Zoologique et Botanique Africaine 19:305-336.

Navás, L. 1932. Insecta orientalia. X Series. Memorie della Pontifica Accademia de Science dei Nuovi Lincei (2)16:921-949.

Navás, L. 1936. Insectes du Congo Belge. Série IX. Revue Zoologique et Botanique Africaine 28:333-368.

Needham, J. G. 1905. Ephemeridae. Bulletin of the New York State Museum 86:17-62, pl. 4-12.

Needham, J. G. 1909. A peculiar new May fly from Scandaga Park. New York State Museum Bulletin 134:71-75.

Needham, J. G. 1920. African stone-flies and May-flies collected by the American Museum Congo Expedition. Bulletin of the American Museum of Natural History 43:35-40.

Needham, J. G. 1927. The Rocky Mountain species of the mayfly genus *Ephemerella*. Annals of the Entomological Society of America 20:107-117.

Needham, J. G. 1932. Three new American mayflies (Ephemeropt.). Canadian Entomologist 64:273-276.

Needham, J. G. & C. Betten. 1901. Aquatic insects in the Adirondacks. Bulletin of the New York State Museum 47:384-612, 37 pl.

Needham, J. G. & H. E. Murphy. 1924. Neotropical mayflies. Bulletin of the Lloyd Library Number 24, Entomological Series 4:1-79.

Needham, J. G., J. R. Traver & Y.-C. Hsu. 1935. The biology of mayflies. Comstock Publishing Co., New York. xvi+759p.

Newman, E. 1853. Proposed division of Neuroptera into two classes. Zoologist 11(appendix):181-204.

Novikova, E. A. & N. Yu. Kluge. 1987. [Systematics of the genus *Baetis* (Ephemeroptera: Baetidae) with description of a new species from Middle Asia] (in Russian). Vestnik Zoologii 1987:8-19.

Penniket, J. G. 1961. Notes on New Zealand Ephemeroptera. I. The affinities with Chile and Australia, and remarks on *Atalophlebia* Eaton (Leptophlebiidae). New Zealand Entomologist 2(6):1-11.

Penniket, J. G. 1962. Notes on New Zealand Ephemeroptera. III. A new family, genus and species. Records of the Canterbury Museum 7:389-398.

Penniket, J. G. 1966. Notes on New Zealand Ephemeroptera. IV. A new siphlonurid subfamily: Rallidentinae. Records of the Canterbury Museum 8:163-175.

Pescador M. L. & Berner, L. 1981. The mayfly family Baetiscidae (Ephemeroptera). Part II. Biosystematics of the genus *Baetisca*. Transactions of the American Entomological Society 107:163-228.

Pescador, M. L. & W. L. Peters. 1980a. A revision of the genus *Homoeoneuria* (Ephemeroptera: Oligoneuriidae). Transactions of the American Entomological Society 106:357-393.

Pescador, M. L. & W. L. Peters. 1980b. Two new genera of cool-adapted Leptophlebiidae (Ephemeroptera) from southern South America. Annals of the Entomological Society of America 73:332-338.

Pescador, M. L. & W. L. Peters. 1982. Four new genera of Leptophlebiidae (Ephemeroptera: Atalophlebiinae) from southern South America. Aquatic Insects 4:1-19.

Pescador, M. L. & W. L. Peters. 1985. Biosystematics of the genus *Nousia* from southern South America (Ephemeroptera: Leptophlebiidae: Atalophlebiinae). Journal of the Kansas Entomological Society 58:91-123.

Peters, J. G. & M. D. Hubbard. 1977. Authorship of the families Siphlonuridae, Palingeniidae and Potamanthidae. Eatonia 23:1-2.

Peters, W. L. 1969. *Askola froehlichi*, a new genus and species from southern Brazil (Leptophlebiidae: Ephemeroptera). Florida Entomologist 52:253-258.

Peters, W. L. 1971. A revision of the Leptophlebiidae of the West Indies (Ephemeroptera). Smithsonian Contributions to Zoology 62:1-48.

Peters, W. L. 1979. Taxonomic status and phylogeny of *Habrophlebia* and *Habroleptoides* (Leptophlebiidae: Ephemeroptera). Pages 51-56 *in* K. Pasternak & R. Sowa, eds. Proceedings of the 2nd International Conference on Ephemeroptera. Państwowe Wydawnictwo Naukowe, Warzawa-Kraków.

Peters, W. L. 1980. Phylogeny of the Leptophlebiidae (Ephemeroptera): an introduction. Pages 33-41 *in* J. F. Flannagan & K. E. Marshall, eds. Advances in Ephemeroptera Biology. Plenum Press, New York.

Peters, W. L. 1981. *Coryphorus aquilus*, a new genus and species of Tricorythidae from the Amazon Basin (Ephemeroptera). Aquatic Insects 3:209-217.

Peters, W. L. & G. F. Edmunds, Jr. 1964. A revision of the generic classification of the Ethiopian Leptophlebiidae (Ephemeroptera). Transactions of the Royal Entomological Society of London 116:225-253.

Peters, W. L. & G. F. Edmunds, Jr. 1970. Revision of the generic classification of the Eastern Hemisphere Leptophlebiidae. Pacific Insects 12:157-240.

Peters, W. L. & G. F. Edmunds, Jr. 1972. A revision of the generic classification of certain Leptophlebiidae from southern South America (Ephemeroptera). Annals of the Entomological Society of America 65:1398-1414.

Peters, W. L. & G. F. Edmunds, Jr. 1990. A new genus and species of Leptophlebiidae: Atalophlebiinae from the Celebes (Sulawesi) (Ephemeroptera). Pages 327-335 *in* I. C. Campbell, ed. Mayflies and Stoneflies: Life Histories and Biology, Kluwer Academic Publishers, Dordrecht.

Peters, W. L., M. T. Gillies & G. F. Edmunds, Jr. 1964. Two new genera of mayflies from the Ethiopian and Oriental Regions (Ephemeroptera: Leptophlebiidae). Proceedings of the Royal Entomological Society of London (B)33:117-124.

Peters, W. L. & J. G. Peters. 1980. The Leptophlebiidae of New Caledonia (Ephemeroptera). Part II. Systematics. Cahiers O.R.S.T.O.M., Series Hydrobiologie 13:61-82.

Peters, W. L. & J. G. Peters. 1981a. The Leptophlebiidae: Atalophlebiinae of New Caledonia (Ephemeroptera). Part III. - Systematics. Revue d'Hydrobiologie Tropical 14:233-243.

Peters, W. L. & J. G. Peters, 1981b. The Leptophlebiidae: Atalophlebiinae of New Caledonia (Ephemeroptera). Part IV. - Systematics. Revue d'Hydrobiologie Tropical 14:245-252.

Peters, W. L., J. G. Peters & G. F. Edmunds, Jr. 1978. The Leptophlebiidae of New Caledonia (Ephemeroptera). Part I. Introduction and systematics. Cahiers O.R.S.T.O.M., Séries Hydrobiologie 12:97-117.

Phillips, J. S. 1930a. A revision of New Zealand Ephemeroptera. Part I. Transactions of the New Zealand Institute 61:271-334.

Phillips, J. S. 1930b. A revision of New Zealand Ephemeroptera. Part 2. Transactions of the New Zealand Institute 61:335-390.

Pictet, F. J. 1843-1845. Histoire naturelle générale et particulière des insectes névroptères. Famille des éphémérines. Chez J. Kessmann et Ab. Cherbuliz, Geneva. 300 pp, xix + 47 pl.

Pierce, W. D. 1945. Two new fossils from the Upper Miocene of the Puente Hills. Bulletin of the Southern California Academy of Sciences 44:3-6.

Ping, C. 1935. On four fossil insects from Sinkiang. Chinese Journal of Zoology 1:107-115.

Puthz, V. 1975. Eine neue Caenidengattung aus dem Amazonasgebiet (Insecta: Ephemeroptera: Caenidae). Amazoniana 5:411-415.

Riek, E. F. 1955. Revision of the Australian mayflies (Ephemeroptera). I. Subfamily Siphlonurinae. Australian Journal of Zoology 3:266-280.

Riek, E. F. 1963. An Australian mayfly of the family Ephemerellidae (Ephemeroptera). Journal of the Entomological Society of Queensland 2:48-50.

Riek, E. F. 1973. The classification of the Ephemeroptera. Pages 160-178 in W. L. Peters & J. G. Peters (eds.), Proceedings of the First International Conference on Ephemeroptera. E. J. Brill, Leiden.

Riek, E. F. 1976. An unusual mayfly (Insecta: Ephemeroptera) from the Triassic of South Africa. Paleontologica Africana 19:149-151.

Rohdendorf, B. B. 1957. Paleoentomologicheskie Issledovaniya v SSSR. Trudy Paleontologicheskogo Instituta 66.

Savage, H. M. 1982. A curious new genus and species of Atalophlebiinae (Ephemeroptera: Leptophlebiidae) from the southern coastal mountains of Brazil. Studies on Neotropical Fauna and Environment 17:209-217.

Savage, H. M. 1983. Perissophlebiodes, a replacement name for Perissophlebia Savage nec Tillyard (Ephemeroptera: Leptophlebiidae). Entomological News 94:204.

Savage, H. M. & W. L. Peters. 1978. Fittkaulus maculatus a new genus and species from northern Brazil (Leptophlebiidae: Ephemeroptera). Acta Amazonica 8:293-298.

Savage, H. M. & W. L. Peters. 1983. Systematics of Miroculis and related genera from northern South America (Ephemeroptera: Leptophlebiidae). Transactions of the American Entomological Society 108:491-600.

Schoenemund, E. 1929. Habroleptoides, eine neue Ephemeropteren-Gattung. Zoologischer Anzeiger 80:222-232.

Scudder, S. H. 1880. The Devonian Insects of New Brunswick. Anniversary Memoires of the Boston Society of Natural History, 41p, 1 pl.

Sellards, E. H. 1907. Types of Permian Insects. Part II. Plectoptera. American Journal of Science (4)23:345-355.

Sellards, E. H. 1909. Types of Permian Insects. Part III. Megasecoptera, Oryctoblattinidae and Protorthoptera. American Journal of Science (4)27:151-173.

Selys-Longchamps, E. de. 1888. Catalogue raisonné des orthoptères et des Névroptères de Belgique. Annales de la Société Entomologique de Belgique 32:103-203.

Sinitshenkova, N. D. 1976. [New early Cretaceus mayflies (Insecta, Ephemeroptera) from eastern Transbaikal] (in Russian). Paleontologicheskii Zhurnal 1976(2):85-93, pl. 6.

Sinitshenkova, N. D. 1985. [Jurassic mayflies (Ephemerida-Ephemeroptera) of southern Siberia and western Mongolia] (in Russian). Jurskie Nasekomie Sibiri i Mongolii. Trudi Paleontologicheskogo Instituta 211:11-23, pl. 1-2.

Sinitshenkova, N. D. 1986. [Mayflies. Ephemerida (=Ephemeroptera)] (in Russian). "Nacekomie v Rannemelovikh Ekosistmakh Zapadnoi Mongolii". Trudy Sovmestnaya Sovetsko-Mongolskaya Paleontologicheskaya Ekspeditsiya 28:45-46, pl. 1 fig. 1-2.

Sinitshenkova, N. D. 1989. [New Mesozoic mayflies (Ephemerida) from Mongolia] (in Russian). Paleontologicheskii Zhurnal 1989(3):30-41, pl. 2.

Sivaramakrishnan, K. G. 1984. A new genus and species of Leptophlebiidae: Atalophlebiinae from southern India (Ephemeroptera). International Journal of Entomology 26:194-203.

Sivaramakrishnan, K. G. 1985. New genus and species of Atalophlebiinae (Ephemeroptera: Leptophlebiidae) from southern India. Annals of the Entomological Society of America 78:235-239.

Soldán, T. 1978. New genera and species of Caenidae (Ephemeroptera) from Iran, India and Australia. Acta Entomologica Bohemoslavaca 75:119-129.

Soldán, T. 1986. A revision of the Caenidae with ocellar tubercles in the nymphal stage (Ephemeroptera). Acta Universitatis Carolinae - Biologica 1982-1984:289-362.

Spieth, H. T. 1940. The genus *Ephoron*. Canadian Entomologist 72:109-111.

Spieth, H. T. 1941. Taxonomic studies on the Ephemeroptera. II. The genus *Hexagenia*. American Midland Naturalist 26:233-280.

Stephens, J. F. 1835. Illustrations of British Entomology, Mandibulata 6:53-70, pl. 29.

Sukatskene, I. K. 1962. [Mayflies (Ephemeroptera) of the Angara River and its tributaries in the area of the Bratsk Hydro-electric Station Reservoir] (in Russian). Lietuvos TSR Mokslu Akademijos Darbai, C Serija, 2:107-122.

Suter, P. J. 1984. A redescription of the genus *Tasmanocoenis* Lestage (Ephemeroptera: Caenidae) from Australia. Transactions of the Royal Society of South Australia 108:105-111.

Thenius, E. 1979. Lebensspuren von Ephemeropteren-Larven aus dem Jung-Tertiär des Wiener Beckens. Annalen der Naturhistorisches Museum Wien 82:177-188.

Thew, T. B. 1960. Revision of the genera of the family Caenidae (Ephemeroptera). Transactions of the American Entomological Society 86:187-205.

Tiensuu, L. 1935. On the Ephemeroptera-fauna of Laatokan Karjala (Karelia Ladogensis). Suomen Hyönteistieteellinen Aikakauskirja 1:3-23.

Tillyard, R. J. 1921. A new genus and species of May-fly (Order Plectoptera) from Tasmania, belonging to the family Siphlonuridae. Proceedings of the Linnaean Society of New South Wales 46:409-412, pl. 34.

Tillyard, R. J. 1932. Kansas Permian Insects. Part 15. The order Plectoptera. American Journal of Science (5)23:97-135, 237-272.

Tillyard, R. J. 1933. The mayflies of the Mount Kosciusko region. I. (Plectoptera). Introduction and family Siphlonuridae. Proceedings of the Linnaean Society of New South Wales 58:1-32, pl. 1.

Tillyard, R. J. 1936. Kansas Permian Insects. Part 16. The order Plectoptera (contd.): The family Doteridae, with a note on the affinities of the order Protohymenoptera. American Journal of Science (5)32:435-453.

Towns, D. R. 1983. A revision of the genus *Zephlebia* (Ephemeroptera: Leptophlebiidae). New Zealand Journal of Zoology 10:1-52.

Towns, D. R. & W. L. Peters. 1979a. Three new genera of Leptophlebiidae (Ephemeroptera) from New Zealand. New Zealand Journal of Zoology 6:213-235.

Towns, D. R. & W. L. Peters. 1979b. New genera and species of Leptophlebiidae (Ephemeroptera) from New Zealand. New Zealand Journal of Zoology 6:439-452.

Traver, J. R. 1931. A new mayfly genus from North Carolina. Canadian Entomologist 63:103-109.

Traver, J. R. 1932. *Neocloeon*, a new mayfly genus (Ephemerida). Journal of the New York Entomological Society 40:365-372, pl. 14.

Traver, J. R. 1933. Heptagenine mayflies of North America. Journal of the New York Entomological Society 41:105-125.

Traver, J. R. 1935. Two new genera of North American Heptageniidae (Ephemerida). Canadian Entomologist 67:31-38.

Traver, J. R. 1938. Mayflies of Puerto Rico. Journal of Agriculture of the University of Puerto Rico 22:5-42, pl. 1-3.

Traver, J. R. 1939. Himalayan mayflies (Ephemeroptera). Annals and Magazine of Natural History (11)4:32-56.

Traver, J. R. 1956. A new genus of Neotropical mayflies (Ephemeroptera, Leptophlebiidae). Proceedings of the Entomological Society of Washington 58:1-13.

Traver, J. R. 1958. The subfamily Leptohyphinae (Ephemeroptera: Tricorythidae). Part I. Annals of the Entomological Society of America 51:491-503.

Traver, J. R. 1959. Uruguayan mayflies. Family Leptophlebiidae: Part I. Revista de la Sociedad Uruguaya de Entomologia 3:1-13, pl. 1-3.

Traver, J. R. & G. F. Edmunds, Jr. 1968. A revision of the Baetidae with spatulate-clawed nymphs (Ephemeroptera). Pacific Insects 10:629-677.

Tshernova, O. A. 1934. [A new widely distributed Ephemeroptera genus from the northern region of the USSR] (in Russian). Doklady Akademii Nauk SSSR 4:240-243.

Tshernova, O. A. 1948. [On a new genus and species of mayfly from the Amur Basin (Ephemeroptera, Ametropodidae)] (in Russian). Doklady Akademii Nauk SSSR 60(8):1453-1855.

Tshernova, O. A. 1961. [On the taxonomic position and geological age of the genus *Ephemeropsis* Eichwald (Ephemeroptera, Hexagenitidae)] (in Russian). Entomologischeskoe Obozrenie 40:858-869.

Tshernova, O. A. 1962. [Ephemeroptera larvae from the Neogene of West Siberia (Ephemeroptera, Heptageniidae)] (in Russian). Zoologicheskii Zhurnal 41:943-945.

Tshernova, O. A. 1967. [Mayfly of the Recent family in Jurassic deposits of Transbaikalia (Ephemeroptera, Siphlonuridae)] (in Russian). Entomologicheskoe Obozrenie 46:322-326.

Tshernova, O. A. 1968. [New mayfly from Karatau (Ephemeroptera)] (in Russian). Pages 23-25, 1 pl., in Yurskie Nasekomie Karatau, Izdatelstro "Nauka", Moskva.

Tshernova, O. A. 1969. [New early Jurassic May-flies (Ephemeroptera, Epeoromimidae, Mesonetidae)] (in Russian). Entomologicheskoe Obozrenie 48:153-161, 1 pl.

Tshernova, O. A. 1971. [May-fly from fossil pitch of Polar Siberia (Ephemeroptera, Leptophlebiidae)] (in Russian). Entomologicheskoe Obozrenie 50:612-618.

Tshernova, O. A. 1972. [Some new species of mayflies from Asia (Ephemeroptera, Heptageniidae, Ephemerellidae)] (in Russian). Entomologicheskoe Obozrenie 51:604-614.

Tshernova, O. A. 1974. [The generic composition of the mayflies of the family Heptageniidae (Ephemeroptera) in the Holarctic and Oriental Region] (in Russian). Entomologicheskoe Obozrenie 53:801-814.

Tshernova, O. A. 1977. [Distinctive new mayfly nymphs (Ephemeroptera; Palingeniidae, Behningiidae) from the Jurassic of Transbaikal] (in Russian). Paleontologicheskii Zhurnal 1977:91-96, 1 pl.

Tshernova, O. A. 1978. [Systematic position of the genus *Paracinygmula* Bajkova, 1975 (Ephemeroptera, Heptageniidae)] (in Russian). Entomolgicheskoe Obozrenie 57:540-542.

Tshernova, O. A. 1981. [On the systematics of adult mayflies of the genus *Epeorus* Eaton, 1881 (Ephemeroptera, Heptageniidae)] (in Russian). Entomolgicheskoe Obozrenie 60:323-336.

Tshernova, O. A. & O. Ya. Bajkova. 1960. [On a new genus of mayflies (Ephemeroptera, Behningiidae)] (in Russian). Entomolgicheskoe Obozrenie 39:410-416.

Tshernova, O. A. & V. V. Belov. 1982. Systematic position and synonymy of *Cinygma tibiale* Ulmer 1920 (Ephemeroptera, Heptageniidae). Entomologischen Mitteilungen aus dem Hamburgischen Zoologischen Museum und Institut 7:193-194.

Tshernova, O. A., N. Yu. Kluge, N. D. Sinitshenkova & V. V. Belov. 1986. [Order Ephemeroptera - Mayflies] (in Russian). Keys to the insects of Far Eastern USSR in six volumes, Leningrad. 1:99-142.

Tshernova, O. A. & N. D. Sinitshenkova. 1974. [A new fossil genus and species of the mayfly family Hexagenitidae (Ephemeroptera) from the South of the European part of the USSR and its connection with Recent mayflies] (in Russian). Entomologicheskoe Obozrenie 53:130-136, 1 pl.

Uéno, M. 1931. Contribution to the knowledge of Japanese Ephemeroptera. Annotationes Zoologicae Japoneses 13:189-226, pl. 12-13.

Uéno, M. 1969. Mayflies (Ephemeroptera) from various regions of Southeast Asia. Oriental Insects 3:221-238.

Ulmer, G. 1914. Ordn. Ephemeroptera (Agnátha), Eintagsfliegen. Pages 95-99 *in* P. Brohmer, Fauna von Deutschland.

Ulmer, G. 1920a. Übersicht über die Gattungen der Ephemeropteren, nebst Bemerkungen über einzelne Arten. Stettiner Entomologische Zeitung 81:97-144.

Ulmer, G. 1920b. Neue Ephemeropteren. Archiv für Naturgeschichte 85(A):1-80.

Ulmer, G. 1924a. Einige alte und neue Ephemeropteren. Konowia 3:23-37.

Ulmer, G. 1924b. Trichopteren und Ephemeropteren. Denkschriften, Mathematisch-Naturwissenschaftliche Klasse, Akademie der Wissenschaften in Wien 99:1-9.

Ulmer, G. 1932. Bemerkungen über die Seit 1920 neu aufgestellten Gattungen der Ephemeropteren. Stettiner Entomologische Zeitung 93:204-219.

Ulmer, G. 1939-1940. Eintagsfliegen (Ephemeropteren) von den Sunda-
Inseln. Archiv für Hydrobiologie, Supplement 16:443-692, fig. 1-
469, 4 tab.

Ulmer, G. 1943. Alte und neue Eintagsfliegen (Ephemeropteren) aus
Süd- und Mittleamerika. Stettiner Entomologische Zeitung
104:14-46.

Walsh, B. D. 1863a. Observations on certain N. A. Neuroptera, by H.
Hagen, M. D., of Koenigsberg, Prussia; translated from the
original French ms., and published by permission of the author,
with notes and descriptions of about twenty new N. A. species of
Pseudoneuroptera. Proceedings of the Entomological Society of
Philadelphia 2:167-272.

Walsh, B. D. 1863b. List of the Pseudoneuroptera of Illinois contained in
the cabinet of the writer, with descriptions of over forty new
species, and notes on their structural affinities. Proceedings of
the Academy of Natural Sciences of Philadelphia 1862:361-402.

Waltz, R. D. & W. P. McCafferty. 1985. *Moribaetis*: A new genus of
Neotropical Baetidae (Ephemeroptera). Proceedings of the
Entomological Society of Washington 87:239-251.

Waltz, R. D. & W. P. McCafferty. 1987a. New genera of Baetidae
(Ephemeroptera) from Africa. Proceedings of the Entomological
Society of Washington 89:95-99.

Waltz, R. D. & W. P. McCafferty. 1987b. Generic revision of *Cloeodes* and
description of two new genera (Ephemeroptera: Baetidae).
Proceedings of the Entomological Society of Washington 89:177-
184.

Waltz, R. D. & W. P. McCafferty. 1987c. New genera of Baetidae for
some Nearctic species previously included in *Baetis* Leach
(Ephemeroptera). Annals of the Entomological Society of America
80:667-670.

Waltz, R. D. & W. P. McCafferty. 1987d. Systematics of *Pseudocloeon*,
*Acentrella*, *Baetiella*, and *Liebebiella*, new genus (Ephemeroptera:
Baetidae). Journal of the New York Entomological Society
95:553-568.

Waltz, R. D., W. P. McCafferty & J. H. Kennedy. 1985. *Barbaetis*: A new
genus of eastern Nearctic mayflies (Ephemeroptera: Baetidae).
Great Lakes Entomologist 18:161-165.

Westwood, J. O. 1840. An introduction to the modern classification of
insects. Longman, Orme, Brown, Green, and Longmans, London.
2 vol.

Whalley, P. E. S. & E. A. Jarzembowski. 1985. Fossil insects from the
lithographic limestone of Montsech (Late Jurassic -Early Cretaceous), Lérida Province, Spain. Bulletin of the British Museum
of Natural History (Geology) 38(5):381-412.

Whiting, E. R. & D. M. Lehmkuhl. 1987a. *Raptoheptagenia cruentata*, gen. nov. (Ephemeroptera: Heptageniidae), new association of the larvae previously thought to be *Anepeorus* with the adult of *Heptagenia cruentata* Walsh. Canadian Entomologist 119:405-407.

Whiting, E. R. & D. M. Lehmkuhl. 1987b. *Acanthomola pubescens*, a new genus and species of Heptageniidae (Ephemeroptera) from western Canada. Canadian Entomologist 119:409-417.

Williamson, [H.] 1802. On the *Ephoron leukon*, usually called the white fly of the Passaick River. Transactions of the American Philosophical Society 5:71-73.

Wu, X.-Y. & D.-S. You. 1986. A new genus and species of Potamanthidae from China (Ephemeroptera). Acta Zootaxonomica Sinica 11:401-405.

Zurwerra, A. & I. Tomka. 1985. *Electrogena* gen. nov., eine neue Gattung der Heptageniidae (Ephemeroptera). Entomologische Berichte Luzern nr. 13:99-104.